As in the Heart,
So in the Earth

As in the Heart, So in the Earth

Reversing the Desertification of the Soul and the Soil

Pierre Rabhi

Foreword by Yehudi Menuhin
Translated by Joseph Rowe

Park Street Press
Rochester, Vermont

Park Street Press
One Park Street
Rochester, Vermont 05767
www.ParkStPress.com

Park Street Press is a division of Inner Traditions International

Originally published in French under the title *Parole de Terre* by Albin Michel, S.A., Paris
First U.S. edition published in 2006 by Park Street Press

LIBRARY OF CONGRESS CATALOGING-IN-PUBLICATION DATA

Rabhi, Pierre, 1936–
    [Parole de terre. English]
    As in the heart, so in the earth : reversing the desertification of the soul and the soil /
Pierre Rabhi ; foreword by Yehudi Menuhin ; translated by Joseph Rowe.—1st U.S. ed.
      p. cm.
    ISBN-10: 1-59477-081-6 (pbk.)
    ISBN-13: 978-1-59477-081-4
    1. Human ecology. 2. Environmental ethics. 3. Spirituality. I. Title.
    GF80.R33 2006
    304.2—dc22

                                                            2006003356

Printed and bound in the United States

10   9   8   7   6   5

Text design by Priscilla Baker
Text layout by Virginia Scott Bowman
This book was typeset in Sabon and Agenda with Bell as the display typeface

To send correspondence to the author of this book, mail a first-class letter to the author c/o Inner Traditions • Bear & Company, One Park Street, Rochester, VT 05767, and we will forward the communication.

Inner Traditions wishes to express its appreciation for assistance given by the government of France through the National Book Office of the Ministère de la Culture in the preparation of this translation.

Nous tenons à exprimer nos plus vifs remerciements au government de la France et le ministère de la Culture, Centre National du Livre, pour leur concours dans le préparation de la traduction de cet ouvrage.

Contents

Foreword

Pierre Rabhi is calling us urgently toward an act of reconciliation with the earth, and his own life is an exemplary model of it. The action he calls for is both real and symbolic, profoundly practical and profoundly sacred. Reconciliation with our Mother Earth is even more urgent than reconciliation with each other, because our very life depends on hers. No life can survive on a barren planet.

With his own hands, Pierre Rabhi has brought life to the desert sands, for life is One, and the fertility of bacterial life transforms the desert itself to renew the life of many species. This simple and saintly man with a clear and truthful mind uses a beautifully poetic language to translate both his burning passion and the determination that has led him to sprinkle barren lands with his own sweat as he successfully labors to restore the chain of life's links, which we are always breaking.

In this book he has chosen an ancient form of storytelling: it is a tale of human arrogance whose desire to control life destroys it, whose desire to control other species leads to their extinction, and whose desire to dominate the earth tortures, mutilates, and profanes it. As an American citizen, I have often been struck by the way the word *dirt* is used in the United States to refer to "earth," whereas the word *earth* is reserved for the name of our planet.

Money has become a universal substance that usurps the place of the earth. In itself, money is merely a tool that represents concrete, living realities and facilitates their exchange. At first glance, it fulfills a practical and useful function—yet the avid amassing of it gives rise to a relentless, insane desire for possession without limit. People are themselves possessed by the illusion that anything can be acquired with enough money—even love, devotion, health, trust, and friendship.

The result is the sickness of whole populations and a climate of permanent aggression, with governments incapable of protecting victims, even when they are neighbors or their own citizens. We become a civilization without self-respect, without respect for different cultures, and without respect for life itself.

This is part of the message of this true and profoundly moving book by Pierre Rabhi.

YEHUDI MENUHIN

Yehudi Menuhin (1916–1999) was a violinist and conductor. Alongside his renown as a great musician, the late Lord Menuhin was equally recognized for his committed humanism. He was awarded the Nehru Peace Prize for international understanding in 1960 and became the Goodwill Ambassador of Unesco in 1992. He founded the International Menuhin Music Academy for young graduate string players in 1977 in Gstaad, Switzerland, the site of the annual Menuhin Music Festival. In 1991, the International Yehudi Menuhin Foundation was established to encourage the expression and representation of cultural identities, especially those that are endangered.

Introduction

The narrative that follows is dedicated to the vast world of oral tradition.

In writing it, I have never ceased to think of those multitudes of innocent beings who have no access to the kind of knowledge conveyed through the written word. Living in a world where exchange of ideas is limited to the spoken word, they lack the means of either participating in world history or forewarning and protecting themselves from its ravages.

My own illiterate grandmother has been especially present to me in this writing. A true daughter of the desert, she was deeply disconcerted by the encroachments of the modern world. She deliberately maintained her own disdainful ignorance of them, as if to exorcise them. Nevertheless, modernity was violently changing her own life and world, and she compensated for her inability to understand it with an intuition that was all the more penetrating and a redoubled affirmation of the values that dwelled in her, as the spirit dwells in the tabernacle. Her essential reproach of modernity was that it abolishes all sense of the sacred, thus exposing humanity to all manner of chaos and transgression.

The first time she smelled gasoline, she said: "This liquid was made by corruption. They had better leave it in the earth, where God put it, or the whole world will be corrupted by it." When she heard the "talking box," as she called the radio, she said: "The Rumis* are magicians, but their miracles are made through pride; they cannot understand real miracles."

Like so many other humans who share her condition—a life centered around survival, a life with neither luxury nor hypocrisy—this woman had become a kind of fortress. Thus she foiled our every attempt to impress or surprise her with the "curiosities" of the new times. Her stubbornness annoyed us, for we were already beginning to be enthralled by the West and its marvels.

Little by little, and unbeknownst to ourselves, we were preparing to relinquish the values that centuries of patience and discipline had wrought in our people. We were preparing to jettison the entire heritage of a bygone era. It was a heritage that caused individuality to ripen slowly, forging the character of a people, building solidarity among them and connecting them to their land. And it was a heritage that, in spite of the many hardships of its people, made them glad to be alive.

People like my grandmother—the "backward" and evermore uprooted peoples of the earth—today number in the billions on our planet. It is as if they were standing on the railway platform of a strange destiny, watching the train of history—a train that has no place for them—roar past without stopping. Little do they realize the extent to which this train is also destroying their lives. It uses their labor and resources and then discards these people, reducing them to human debris. At the very most, it offers them the option of submitting to its domination and all its perversions—and this only on condition that they quickly fall in step with its agenda, or else disappear. This world does not want to wait for them, to understand them or love them. It is too busy hurtling into its own meaninglessness and toward its own demise.

Born of my own imagination, the figure of Tyemoro symbolizes my feelings of love, compassion, and admiration for all the authentic small farmers of the earth. Both in the north and south, these shapers of the clay of which they are made often incarnate its strength and its silence.

[*Rumi in Arabic means "Roman." It is an ancient term that can refer specifically to Europeans (as it does in this instance), or to other inhabitants of lands formerly part of the Roman Empire. —*Trans.*]

Everywhere they live they are now experiencing suffering and betrayal. Although this is an African tale, its intent is deliberately universal. This is why neither the Batifon people nor their land are to be found in any factual records.

After decades of the power of science and technology convincing many humans that they are like gods, disillusionment is growing and becoming a vital force. All the tumultuous frenzy of the twentieth century, with its orgy of homage to materialism threatening all life that is grounded in sensitivity, intuition, and feeling, seems to be moving toward an immense twenty-first century climax of doubt. The world is full of anguish and imbalance. Barbarism is already here, lurking in people's hearts in the same way that cold-blooded massacre lurks in our nuclear arsenals. This world order can congratulate itself on at least one unparalleled achievement: the most superlative and hideous offering ever made to the forces of destruction.

The fictional form of this narrative may be misleading. Its intention is not to entertain but to awaken all consciences to the violation of the earth, so cleverly concealed in current agricultural policies, and, by the irrevocable law that links us to the earth, the violation of ourselves as well.

Yet no initiation into the visible and material ways in which the earth nourishes us must lead us to neglect its implicit spiritual dimension. Our age of rationalistic materialism, weighed down by its obsession with the mineral realm, spreading its toxic substances everywhere, and hypnotized by a science that itself is becoming more and more enslaved to the profit motive, has seriously damaged the vital envelope keeping our planet alive. It seems that only the dimension of the sacred can provide a measure of the awesomeness of our responsibility. By *sacred* I mean a sense of humility in which gratitude, knowledge, wonder, respect, and mystery all come together to inspire and enlighten our actions, transforming us into beings who are truly present in the world and free of the vanity and arrogance that reveal far more our fears and weaknesses than our true strength.

My hope for this book is that it can contribute in some way toward the fundamental need and right of human beings to be able to feed themselves in all the places where that right is being threatened. My hope is also that it will lead to a deeper meditation and reflection upon the fertility of the earth and on our need to establish a new and vital pact with it.

This tale grows out of an experience that is both objective and personal, both practical and vital. It involves issues that affect every human being, for it concerns our Mother Earth, the source of our nourishment, our life, and our survival.

1
Return to Membele

The small, old truck that served as bush taxi labored mightily, devouring mile after mile of rough and jolting road creviced by the violence of the rare downpours and eroded by the relentless winds of the dry season. In its wake, the taxi stirred up a cloud of ochre dust that mingled with our own sweat, making it stick to our faces. There was a suffocating heat as well, filled with smells that evoked many memories for me. My anonymous companions and I were jammed together, becoming a single mass following the movements of the overloaded taxi, lurching and swaying as if in a state of common drunkenness. From the constant roar of the vehicle there emerged somehow the sound of a voice singing nonchalantly. I finally discerned its source: it came from somewhere behind the long scarves wrapped around every face in protection from the dust. Only our eyes were visible, and a brief exchange of glances—barely substantial, but enough to express simple feelings of mutual recognition and goodwill—was our only means of establishing tentative communication. A few of the travelers began to speak to each other in loud voices. They were Batifons, and they would never have imagined that this lone white foreigner sitting among them could understand everything they were saying. They would have been even more astonished to know that I had actually written about their language—its origins, structure, and symbols—as well as about the people who speak it. Batifon is my specialty as an ethnologist.

I have long considered my discipline to be a voyage of initiation lasting as long as life itself. Beyond the last breaths of the dying elders, the initiation continues through other minds and other memories. It was seemingly by chance that I found myself in this particular current of tradition, transmitted from generation to generation by elders who anticipated the day when they would no longer be able to contribute to its weaving, who are now anxious to protect it from decline when its fabric begins to unravel.

The Batifon community is one of those now threatened with extinction. At the same time that the modern world of money and commerce wreaks this kind of destruction, it also produces specialists like myself, who try to salvage some vestiges of these dying cultures. My work is both shameful and exalting, both a grief and a privilege. My predecessors— missionaries, explorers, anthropologists, and traders—have often functioned as game flushers, driving into the voracious nets of destruction hidden behind all its hypocritical pretensions of "civilization" all those hapless peoples who never asked anything of the West. In spite of all the humanitarian pretexts, which have made excellent rationalizations, the invaders have mostly been unable to recognize these "savages" as members of their own species.

Now well advanced, these cultural devastations would have been less painful if the modern world and its progress had offered any credibility as a model. But its own "development" turned out to be ambivalent at best. The miracles accomplished with the help of its mathematics and material science have not been enough, especially when we take into account the destructive power of these "miracles." For example, the perfection and propagation of armaments is actually causing humanity to seriously regress almost in the metaphysical sense of the term, for we have profaned death itself. Death has become strangely banalized, and some of the most ignoble acts become mere details of journalism and history. What is worse, we even honor these acts. And this not our only failing.

What is the meaning of modernity? Its characteristic trait is its glorification of inanimate matter, from coal and steel to petroleum and the atom. And what will it offer tomorrow? More of these miracles? But who and what do they serve? Is all this "progress" worth the sacrifice of the myriad innocent peoples and cultures that, not so long ago, were spread all over the globe, like germinating seeds?

As the bush taxi brought me closer to my friend Tyemoro, my head was heavy with these somber thoughts. This, my fifth field trip, had a different feeling from all the others. As an informant, Tyemoro had already given me a wealth of material, including explanations that greatly advanced my analysis of the language he mastered so beautifully, as well as the symbolism underlying it. I had enough material to fill hundreds of pages, allowing me to continue to publish and maintain my teaching position at the university. But the most precious gift Tyemoro had given me was that of his own friendship. This old man, imbued with a dignity surpassing that of a prince, was generous and patient; he was one of those rare beings whose inner character has remained whole and authentic. Just being in his presence made my own problems as professor, husband, and father dissipate like smoke in the breeze. When I was with him, I was able to completely forget the world of competition and stress in which my profession requires me to live. To tell the truth, this fifth trip was not really necessary for my formal research, but it was a trip I knew I needed to make—and in spite of the heat, the dust, and my own ruminations, I was already beginning to feel its benefits.

With my first faltering steps on the ground, dizzy and tired from the long trip, I felt I was reestablishing my own connection with this land that had become so familiar to me. The truck sat still and empty. It looked to me as though it was exhausted, panting through all its valves and pipes. I felt a sense of gratitude toward it, for my own fate had been in its charge during this long trip. It was as if the old truck had fulfilled some moral contract to deliver me safe and sound to my destination.

A small crowd quickly gathered around me, for I had been recognized. A sound of joyful greeting arose and, with cries of "François! François!" innumerable hands reached out to grasp mine. My bags were whisked away from me, and an escort formed to lead me to the village and the house of Tyemoro. I would have preferred to arrive less covered with dust, but this was not a community that stood on formalities; everything was allowed to happen in its own good time. With my escort,

I walked between the adobe houses. I was now completely captivated by the atmosphere: the sounds, the smells, even the dust, all conspired to reintegrate me into this other world that had become my own.

Approaching Tyemoro's house, I also felt as if I were a son returning home. My admiration for him was mixed with a profound mutual affection. As soon as we arrived in his courtyard, I saw him standing there in front of his room, dressed in a robe the color of sand, so long it reached to his heels. He had hardly changed in the last three years, with his white hair and beard in striking contrast to his handsome, dark brown face. Before touching me, he paused to peer at me, his eyes sparkling with life and intelligence in spite of the white spots that partially marred them. Finally, he held his arms wide open to me, and my whole body was received in an embrace whose warmth and strength was a reflection of the length of time we had not seen each other.

Having bathed and then dressed in the light and comfortable clothes of local custom, my body itself felt lighter, almost ethereal. Once again, I was living in this other world of mine. It was a strictly masculine world; the women who prepared my food were always enveloped in an aura of discretion and reserve. They came and went with a presence that was so light it was almost ghostly, yet paradoxically was as solid as the roots and trunk of a tree filled with the sap of an ever-renewed courage and tenacity. I had long realized that these women were the foundation of the health and survival of this people. It was they who held together the very framework and life of the community. Yet their world was barely accessible to me, for my dealings were with only the men. To me, the women's world was nourishing and hospitable, yet also closed and secret. They avoided looking directly at me, and it was only through long and subtle observation that I finally discerned a faint, tremulous movement in their expressions, revealing that their indifference to me was feigned. In this culture, words are the domain of men. A woman would be viewed with great disapproval and would even be disgraced if she presumed to engage in unnecessary conversation with a foreign man. I had to be careful and prudently reserved with them so as not to behave clumsily and perhaps upset this order. In spite of appearances, these well-established customs sustained a certain climate of serenity. Thus all I really know of Batifon women is what I have learned from Tyemoro as he described to me the ways of his social system. Contrary to what we might suppose, however, Batifon mythology, unlike ours, in no way denigrates the feminine principle. Woman's reserve and self-effacement are believed to arise from her closeness to the original principle of life. Both seed and seed-bearer, woman represents the realm where action and silence are one.

I knew, though, that in intimate situations these women express themselves fully and with great vehemence, if necessary. The result is that the men gradually become (consciously or unconsciously) the vehicles of their women's opinions, intuitions, and certainties. It is only in this oblique way that they influence the community debate.

With free time on my hands before I meet with Tyemoro, I play the role of a sort of paramedic, distributing medications, disinfecting wounds, putting drops in eyes, and so forth. This is one of the rare occasions when I can be close to the women—they bring me children afflicted with various ills, some of which I can help and others about which I can do nothing. Other than our minimal exchange of words, I content myself with a few shy looks of recognition, and gratitude.

The first days pass and my only job at present is simply to live in the village, allowing time for me to accomplish the work of physical, psychic, and spiritual readaptation. I see Tyemoro every day, often eating lunch or dinner with him, sitting cross-legged on mats that cover the hard earthen floor. These meetings occasionally provide opportunities for some conversation.

I am aware of the extreme fragility of my hosts' material situation, and ignoring their protests, I insist on contributing well beyond their expense of feeding me. Recently a shop has opened here, run by a stranger with light-colored skin. The merchandise he offers seems almost comical in this context: a few tins of sardines and mackerel, tomato sauce, soap, tea, batteries, dried macaroni, rice, and beans. These are exchanged for the rare bills or coins that somehow happen to show up in this isolated place. For many families, every day is another difficult step in the struggle to survive. The very nature of this tiny store is a reflection of the insecurity and precariousness in which these people have been forced to live. Modern life has marked the place with a few of its most elementary products, such as cast-off clothes. Their incongruous presence only accentuates the extent to which the lives of these people are largely centered around survival. Every day is a heroic effort for them. The sand—unknown here a few decades ago—is gradually covering everything, adding to its vast empire. Much of the soil is being choked by it. Malnutrition has marked the most recent generations of children. They can survive longer with the new medicines that arrive, but they lack the solid constitutions of their parents. This forces me to ask a terrible question: Is helping these sick children to survive really saving them or merely condemning them to a longer agony? From poverty to misery, from misery to starvation and death—the process is now so well-established that it has become banal. I am overcome by a feeling of rage against human selfishness and am humiliated by my own powerlessness to do anything about it. I know from my research that the African continent is not lacking in resources, nor is overpopulation its major problem—in fact, it could thrive with a much higher population. Those who broadcast and receive only the tragic images of Africa are ignorant of the valiant efforts made by its men and women to overcome the doom and chaos created by the developments of its recent history.

Living with such people is disconcerting—not least when their condition, which for me inspires grave reflection and a kind of petrified sadness, becomes for them an occasion for bursting out in laughter, even rhythmic handclapping, with a childlike exuberance. This is a kind of miracle in itself, a joyous defiance of the fate that is overtaking them, a hope against hope for a better time that is more just and constructive.

My days here are ordinary, with no surprises. I get up at dawn and walk for a couple of hours in the bush. From a distance, I look back on a scene now animated by the lines of women beginning their tasks—carrying water or gathering wood—and by the men herding their family's livestock. After my walk, I do some reading related to my work or write letters and notes, sometimes reflections and observations, as the mood takes me. I am under no constraints and the time is very agreeable to me. After the meal, drowsiness overcomes me and I take a nap. And then my favorite time of day approaches, when the twilight makes everything grow quiet, contemplative. On moonless nights the whole village disappears. Only a few faint lamps can be seen here and there, striving to push aside the dense curtain of night. Tyemoro's courtyard remains open to everyone. People arrive there and depart quietly in the darkness.

Sometimes I can identify them by their voices, but they speak very softly; verbal exchanges take place as if in a sanctuary. They come just to be in the presence of the old man, without disturbing the silence he favors.

I have not yet begun my series of interviews with him as informant for my research. To tell the truth, I have been putting it off without any clear decision to do so. I have been savoring a kind of carefree mood, giving myself ample time to relax and shed the last vestiges of my other world that stick to me like mud sticking to the soles of my boots. Tyemoro has always left it up to me to decide when to begin and what to talk about. At the beginning of our collaboration, I was paralyzed by a kind of exaggerated caution that made my words and gestures awkward and inconsistent. Perceiving my embarrassment, Tyemoro said a few simple words to put me at ease with him.

2
Ninou's Plea

This routine was suddenly interrupted by an apparently insig-
nificant event, yet one which immediately widened the scope of
my reflections. I was already in bed, my eyes open in the darkness, my
mind occupied with those vague thoughts and images that precede sleep.
Something was moving in the corner of my room. At first I thought it
might be an animal, but then a child's voice spoke: "François, I came to
see you because I can't sleep."

I immediately recognized the voice of little Ninou. He had adopted
me as his special friend and followed me around with affectionate devotion. Nothing made him so happy as to perform some service for me,
bringing me a drink of water or running an errand. The slightest request
of mine caused him to be overcome with joy, and his face would break
into a huge smile, with sparkling eyes that betrayed a kind of comical
pride. I had learned that Ninou, already orphaned at his young age, had
also recently lost his aunt. People tended to consider him simpleminded,
but I felt that he was mostly lacking in affection. It was natural for him
to look for attention from the unknown stranger among them.

His presence in my room had made me wide awake and curious.
Little by little, the darkness seemed to give a kind of grave presence to
the sound of his child's voice. I sat up without speaking, leaving the initiative to him. I sensed that he was sitting next to the door. A long moment
passed in which various questions ran through my mind, yet I could

not articulate them. Finally his voice began speaking a long, meditative monologue. It was astounding and disconcerting, especially coming from a boy who was barely approaching puberty. With every fiber of my body I listened carefully to the strange tale this child had decided to tell, trying in the telling to convey some obscure message to me.

"The old woman was always looking everywhere, all around her. Every day, she walked out of her house, sat on the ground, and looked all around, waiting for something to happen on the road or on the mountain up there, far away. Some days you can't see the mountain, because the dust covers everything. The wind blows, the dust comes up, and you can't even see the sun. Some days the heat is bad, and then the winter comes and it gets cold. But old Meka walked out of her house every day, looking at the desert. When there's no dust, you can see the skinny trees. Maybe God didn't have enough seeds to grow many of them. You see one over here, another over there. When you look at those trees, you can tell they don't have enough to drink. When we get thirsty, we go get water from the well or from the pond, and the animals go there to drink too. But those poor trees can't walk to find water, and they just stand there, their leaves covered with dust, and sometimes their thorns too. The animals come and eat their leaves and even break their branches. The trees never cry or say anything. But old Meka says that the trees are crying. It's true that she's a little upset in her head. But you have to understand, she had eight children, and they all left for the big village* last year.

"Her husband got old and died, broken by hard work. His hands were hard and callused from using hoes and chop-chops. His whole body was dried up and a little bent, but when he looked at you—well, my friend, you couldn't just stand there, looking in those eyes! His mouth said nothing, but his eyes said, 'Yes, I'm poor! But I'm nobody's servant. . . .' When the company came to build the dam at Dani, the boys all left in a truck to work there. Sometimes they came back, but then they went away for a long time. Then, they came back with some *larzan*.†

"They said that this larzan made them live well, and they sent some back to their parents. It's true, their folks did get some of that larzan.

---

*The Batifon term for the city.
†*Larzan* is the local pronunciation of the French *l'argent,* meaning "money."

The old man gave it to his wife, but his wife gave it back to him. They didn't know what to do with it, so they put it on the windowsill with a stone over it so it wouldn't blow away. One morning the larzan had disappeared. I was there, but they didn't say anything. There wasn't much in their house and the walls were cracking in some places. But the old lady, every night she would sweep the house, shaking out the mats and cleaning everything, like they were expecting guests. After she finished, she sat outside the door and looked at the desert.

"One day her husband told her: 'I'm going to leave you now. God is calling me. I would rather stay with you, but you don't argue with God.' Then old Kafa went to lie down and told his wife to give him her hand. They held hands tightly for a long time, without speaking. Outside, the wind was blowing hard, making sounds like those cane flutes the shepherds play. The whole house was singing with it. I never believed a house could sing, but it's true, that house was really singing. When old Kafa's hand lost its grip on hers, she gave a little cry but did not move. The wind was rattling the copper plate hanging on the wall, but it made only a small noise. The village people came to help prepare Kafa's body for burial. Many old people came to the funeral, because most of the young ones had left. Me, I stayed with the old lady from then on. She was my aunt and I didn't want her to be alone.

"Sometimes one of her boys would come to visit, dressed in new clothes, wearing those black eye hiders.* He had several bracelets that tell you how much time has passed. Other boys came riding put-puts, which made a lot of noise and went very fast. Several times I hid because I was afraid of those put-puts. One day I touched one of them and it burned my hand, worse than when you steal a piece of meat from a boiling pot. Every time they came, the kids would try to get their mother to come to the big village, but she always said no. A few of them, like Tobi, the oldest son, got angry. But the old lady said nothing, and Tobi rode away on his iron donkey. Me, I was always kind of afraid of those boys because of the mean way they spoke. One of them, Sina, slapped me because I accidentally knocked over his talking box. The people inside it stopped talking when it fell down, but they came back when he touched

---

*Sunglasses.

it. I was glad, because he was frowning and very angry at me. He said, 'You're lucky this time.' Then he left. Every time they came, the boys gave their mother some larzan. But she just put it on the windowsill, and it was usually stolen. The boys loved their mother and wondered if she was angry because they had all left for the big village. Old Meka didn't say whether or not she was angry, but she prayed to God to protect her children. And every day, just as she did before, she watched the desert. But now her eyes were almost blind. I cared for her two goats and took them out to pasture and water every day. I didn't want them to eat the leaves from those poor trees, but they had to sometimes. I would shove them along, so that they only took a few leaves from each tree. Dust would fly up with every step I took. I would walk far away, almost to the big mountain, and I could see the house behind, in the distance. It looked like a big, brown buffalo lying in the middle of a huge dry field. Last year she had four goats, but two died. They may have been poisoned by something, but it's true they were getting old. Every evening, I brought the goats back and milked them. My Aunt Meka made curds with the milk. Some days I watered the garden where we grew onions, okra, turnips, beans, and cabbage. There wasn't much water in the well, and we had to wait at least three days in between watering the garden.

"Sometimes Uncle Sarindi would come visit Aunt Meka. They spoke a little, but not very much. When Uncle Sarindi was younger, he had camels that carried salt, wood, and grain. He also brought back news from his travels. People liked to hear him talk, because he had interesting stories to tell. He spoke slowly and would sometimes stop so that people would beg him to continue. Then he would close his eyes and no one even dared to cough. But now Uncle Sarindi's camels are long gone and he has nothing much to say. He sat a long time with his sister, but they spoke very little. My aunt boiled water and made some tea. For a long time, the only sound you could hear was their mouths sipping the tea. She would give me a cup or two as well. Uncle Sarindi put his hand on my head to say hello and also to say goodbye. His hand was warm, and I could feel his ring. When he left, he said, 'Take good care of your old aunt, and do what she says.' I don't know why, but I really liked Uncle Sarindi. He walked very straight, not at all like an old man, and his clothes swung and danced around his body.

"Sometimes I would cry, because I had no children to play with or talk to. In the rainy season, I always dug holes in the ground to plant millet seeds. One year so many grew that we didn't have enough storage for the grain we harvested. Other years, the baby plants would come out, but then the rain would stop and the sun burned them and then the sand covered up the baby plants. Sometimes the plants would grow up, but it was so dry they gave almost no grain. One year my Aunt Meka cried because there was no grain. She cried a long time, drying her eyes on her clothes. Finally, she took some of the larzan her boys gave her and bought some grain from the merchant when he passed by. Nobody in the village had grain that year. But every year, we plant the grain and hope it will rain. All of us look at the sky, and even a tiny cloud makes us feel good. But we never speak to each other about the rain. It is as though people are afraid that if we talk about it, the rain will get angry and not come. The fields are burned by the sun, and the wind from the desert helps the sun to dry up everything.

"The people of the village meet together often, under the big tree. My Aunt Meka asked me to go with her to this meeting. She used her cane with her left hand and my shoulder with her right. We walked very slowly, as though we didn't want to disturb a single pebble on the road. Most of us walk slowly like that anyway, when we don't feel like disturbing anything. It's as though we are saying 'excuse me' to the plants, the donkeys, the rocks, the sand, and even the sky: 'Excuse me, I'm just passing by.' Before, when there were still children around, things were different. But now, this is how it is.

"At the meeting almost all the men were very old. There were a few younger women with children on their backs. Everyone was sitting, but no one spoke for a long time. They were all waiting for Tyemoro to speak first. The big tree spread its branches, trying to make as much shade as it could. Its roots came out of the earth like fingers. Its base looked sort of like a giant hand, grabbing a huge chunk of earth. It was the biggest tree around, and already some of the old people were wondering about what would happen if it died. Some of them said they hoped they would never live to see it die. We all treated this tree as though it was a person—a child, or an elder. Many women let some of their water slosh out when they passed near it. The dust swallowed most of this water, but it did keep the ground around the tree wetter for awhile. No one could say why, but we felt this tree was our oldest ancestor. Some old folks used to say it could talk and that it knew the whole history of our tribe.

"Nowadays, our people don't speak much when they get together like this. The only sounds you hear are a few sighs and the swishing of fans. When they would meet years before, folks would speak a lot more—they spoke about all the young people leaving, about the bad harvests, and about those who were sick. Now, they just lower their heads and sigh every once in awhile. They lower their heads because the wind, the sun, and the drought go on and on, and the young men have all left. When they come back, riding their put-puts, they seem like foreigners here. Sometimes I wanted to leave too, to go to the big village and have my own talking box. Sometimes I cried, being stuck here with only Nana, the simpleton, and Biasine, who can't even walk. But I knew that if I left, my aunt would be all alone, with no one to help her. Also, people told me I was too young to go to the big village by myself.

"One morning, my Aunt Meka didn't get out of bed. I went to see her, and she was trembling a lot. Her face was covered with sweat. She called me to her side, took my hand, and caressed my head. She sent me to find the healer. He came, but he couldn't stop her from trembling. People from the village came to help take care of her over the next few days. But one morning she was dead.

"Now, there's nothing left for me here. There's no reason for me to stay. . . . Why don't you take me with you, François? I'd be like your son and run all your errands for you, and you'd teach me many things."

The voice in the night stopped and the silence grew dense and palpable. In that total darkness, I could turn only to my inner imagination and reflection, searching for a response to this supplication. My first impulse was to extricate myself immediately from this "trap" with a categorical refusal. But then I decided to let the time pass, feeling that time itself would finally offer an honest response that would satisfy both the mind and the heart. I arose and tiptoed toward the boy. I took him in my arms, and he began to shake with silent sobs. I felt in him a deep sadness, close to despair. It was at this point that I had a clear vision of the absurdity of my own situation.

We scholars and experts are so absorbed by our books, our speculations and theories, our references to the past, with its vast historical panoramas that offer us material for crafting our clever garlands of words, that we lose touch with the present, with what is right in front of us. This child's visit, like that of a strange angel of the night, changed my entire frame of mind. I came to a very different decision about the direction and theme of this trip. I realized that I had already accumulated a huge amount of information on the past—Batifon language, traditions, culture—yet I knew virtually nothing of the events of the concrete present that is affecting the lives of these people and their land.

3
Tyemoro's Memory

The very next day I made my first formal visit to Tyemoro. He immediately noted this, pointing out that I was carrying my "word trap," as he called my cassette recorder. He led me into his room, for the heat and the flies were unbearable outside.

The old man sat down in the squatting position he always favored when he wished to concentrate. This posture resembles the fetal position somewhat. Strangely, it only enhanced his dignity and nobility.

I spoke first:

"Tyemoro, my friend, I have come now to ask you to speak of your land. I know that things today are not the way they were before. Why have so many young people left? Why has the village shrunk so much so that mostly old people are left? Why do you have so little food now? Why is the land in such bad shape, and can anything be done about it?"

Obviously startled by my questions, the old man remained silent for a long time. It was an uncomfortable silence for me, because I wondered if my questions somehow offended him. In any case, I consoled myself with the certainty that he would have no difficulty dealing with them, even if that were the case. The Batifon language has many exquisite and very courteous formulas for saying no so as not to insult a guest or violate the rules of hospitality. Finally, when he had overcome his surprise, Tyemoro began to speak with his usual eloquence, becoming the very incarnation of his words.

"I have thought much about this, day and night. Long ago, I asked myself why our life was changing like this. Have we offended the Great Designer in some way? Perhaps we have not shown proper respect for his gifts? Or have our children angered the spirits of the ancestors by leaving the way they have? I do not know. Many times I have spent the night alone up on the mountain. I hoped that in the mountain's secret something would come into my mind to tell me what we have done wrong and how we can make up for it. Sitting in the dark night, with the jackals and serpents around me, I stretched my body out on the naked rock, still hot from the day's burning sun. But no answer came to me, either in waking or in sleep.

"How many dawns have I watched being born, glowing greater and brighter before the sun finally appears? Sitting there on the mountain, I looked at the desert below and remembered what my father's father used to say. In his time, there were trees everywhere—so many of them in some places that they crowded against each other, hiding the sky. Our people ate many fruits and wild animals, which were abundant then. There was plenty of water, too, and the people caught many fish. Our ancestors always cultivated little parcels of land around the trees. Later, people started uprooting the trees, and the parcels became bigger. The forest was still full of dangerous animals in those days, and people had to be careful, especially with their children. There were even villages in the middle of the forest, with high wooden fences around them to keep out the animals. Every year, we had to burn the brush close to the village to drive away the poisonous serpents and other animals. We would trap some of them for our food. . . . That is what my grandfather used to say.

"In my father's time, things changed, but not very much. The parcels became bigger, and we began to trade our harvests with those of other people. We gave others things they lacked in exchange for things we lacked. Sometimes this led to quarrels, but not very serious ones. Some of us were better at hunting and others were better at farming. If somebody knew or learned something, everyone benefited from it. Our tribe was like a single body with many heads, arms, and legs. Every healthy child became an elder and every elder was once a child. The young learned from the old and the old were fed by the young. Nothing seemed to change, yet everything was always changing. Our chiefs guided us, for they had learned to listen to the voice of the secret. They showed us how to organize our life together. Our leaders' task was to keep the body of the people healthy, making sure it did not become sick because of the actions of one or another of us. Of course, things were not perfect. Some things were wrong, as always. Some men or women became jealous of each other. A few even became bitter and wicked with such thoughts and might even use poison against another. But still, life was good in the village. There was much that was good here, and we had many festivals.

"But now, we live in a great absence. I do not understand why we find ourselves faced with this great absence. . . . When I was a child, the rainy season always arrived to water our seeds and make them grow. Only once or twice was there not enough rain. People joked about it, saying that the rain must have overslept. We knew its caprices, its delays, and sometimes its droughts. But it was never that bad, and we had nothing to worry about because we had plenty of grain stored from previous years. There was also plenty of wild game and fruits. One year we might have less food, but then the rain would come, and the next year we would refill our stores of grain. When I was a child, you couldn't travel alone because of the wild animals. The village and the fields were little clearings in the vast forest. Some years more children than usual would die, like excess fruits falling from the tree of life. But others were born, and joy and sadness were woven together into the fabric of our days. I don't know why, but something always told us: be patient, for all of this is the will of God. Our parents and ancestors, both the living and the dead, were linked together by this spirit.

"Sometimes strangers would show up mysteriously on our road. They would wave to us from afar as they came, and we knew this meant their hearts were pure, because their open hands showed their good intentions. We told each other: 'These unknown people must be good for us, for we have done nothing to deserve evil.' We always received them with honor: We milked our animals especially for them and might even sacrifice one so we could offer meat to the visitors. Our guests lacked for nothing, and sometimes they would stay longer with us. We took this as a good sign, for it meant they found goodness here.

"These travelers always told us about their lives. Sometimes we

learned strange things from them and sometimes gained useful knowledge. They were also glad to learn from our ways of doing and thinking.
We exchanged proverbs with them, and sometimes long-lasting links
were established.

"Sometimes people would come from hostile tribes. They never
announced their arrival and always brought evil with them, even death
on some occasions. We would then raid their villages in revenge. But this
was no honor to us, for we were full of hatred and enmity, which always
degrades people.

"One day some white men came. We did not understand their language, but they understood ours. They spoke to us about their god. Some
of them treated our children with their medicines. They worried us a little, but they were accompanied by people of our color, and this reassured
us. But these more familiar people were dressed like the white people
and spoke their language. They told us that the whites only wanted to
help us. These people of our color wanted to reassure us because we had
heard about battles with whites, whose terrible weapons made everyone
afraid. One day a white man came with a fire stick and used it to kill an
antelope. When we saw this, we all went into our houses, afraid to come
out. Standing alone outside, the man laughed and shook his stick in the
air, making it thunder over and over. Afterward, we grew more used to
seeing these people, but they always made us nervous. Some of them
were not afraid of the forest. They said that their god, who told them
to leave their country and come here, would protect them. We saw that
our ways did not please them, and they asked us to change them. They
said that ancestors are not important and that their god was the only real
one. I don't know how it happened, but several of our people went over
to them. They began to dress like them and eat like them. They spoke
their language and learned their medicine.

"Things went along without too much change, but these white people finally explained to us that all our lands belonged to them and that
even we belonged to them. This was clear to us one day when a white
man came with a lot of men of our color who were all dressed in the
same way and who carried those fire sticks that kill. Some of our men
argued with them and were taken away, and we never saw them again.
When a few of our men tried to fight them, the fire sticks killed them.

"After this, fear and resignation—and sometimes shame—came to dwell in our hearts. Many of our children learned the white people's language and their way of seeing the world. It was from this time on that we began to feel more and more different from our children. Our young people spoke to us of countries far away, and they knew more about these countries than they did about our own. They spoke to us of the glory of the white men, their great knowledge, and their great god. We resigned ourselves to being seen by them as ignorant. Every day, our life slipped away from us a little more, like a rope through blood-soaked hands. Our old people began to die in a way they never had before: Their eyes no longer closed on their own truth, but on their own doubt. For the first time ever, some people even died weeping. Another thing started happening then: We began to be embarrassed in front of our children when we followed the ceremonies our ancestors taught us—ceremonies for marriages, funerals, and circumcisions. We still played our drums and the whole village still danced, but it wasn't the same. Finally, our children refused even to participate in these old ways.

"Time passed, and the whites became the masters of our minds. On our land they built houses for their god and other houses for their chiefs. These chiefs acted like little kings, telling us what to do and what not to do. One day, they ordered all male villagers to gather in the central place. The main chief chose the strongest men and had them sent away. He explained to us that they would receive great honor and glory because they were going to fight for their mother country—the country of the whites. This land was far away, but the white chief said we owed it all our happiness. Many of our sons never returned, but those who did told us of terrible, terrible things. We could see that they had changed a great deal. The way they walked, the way they looked, their words, and their souls had changed. A few of them were proud to show the jewelry of battle, which the whites had given them. They said that only the bravest men received these.

"We began to live as if in a storm that never ends, where everything is always being blown away. Nothing was in its proper place anymore. Our signs, our words, our gestures, and our customs had become things to be ridiculed. It's true that the whites also did some good things, even if we could not understand how they did them. Their medicine saved some of our children from death. Our mistrust of them was mixed with gratitude, which made us feel confused.

"We older ones were becoming relics of the past. We were filled with nostalgia for the days when everything followed the order established by our ancestors. They had been very careful to maintain this order, just as you carefully tend a good fire to light the night and keep away the dangerous animals. But we knew that we were giving up the struggle to keep the old ways. We thought that God must somehow be in favor of all these upsetting changes.

"And the changes never ceased. One day a group of whites and blacks came, and we were amazed to see how much alike they were. The blacks acted so much like the whites, and the whites acted so much like the blacks that except for their skin, you couldn't tell them apart. These people were sent by a great king to help us put our life in better order. By this time, our children were already speaking their language, and always talking about the cowries,* the larzan, they use. They said that cowries would bring us great prosperity. It's true that our sons sometimes brought home strange things made by the whites. They had paid for them with the larzan they earned from working for the whites. These new things worried us at first, because we thought they could not have been made by human hands—talking boxes, metal donkeys, bracelets that measure time, and the box trapping the cold light that you can make shine anytime you want. All these things made us more and more silent. If we were to continue to honor our God and our ancestors, it seemed we would have to exclude our own children from our ceremonies, for they were now very far from our beliefs. Some of us said our children have betrayed us, but I don't say that—I think they have just been swept up in a great force that neither they nor we understand. This force cares nothing about our own consent for our children to explore these new ways. Now they have become like whites and we can no longer initiate them into our secrets. But at least one command of the ancestors remains strong and clear: never reveal our secrets to anyone whose heart harbors impure feelings.

"So our group, which had been like a single tree, began to lose its

---

*A cowrie is a type of small seashell widely used as a medium of monetary exchange in pre-colonial Africa.

branches. The sap that had nourished our community was draining away, through an invisible wound. Now we are only a trunk, even a stump, with no branches. Our elders are like stores of grain that no one wants to open or eat. Our children have lost all hunger for the substance we offer them. They have been stolen from us by a new world, which came silently like a thief in the night. Our old people are like eggs without a hen to hatch them, closed and silent in a life with no future. The tears we hold back flow inside us, in useless regret. We have lost our oars, and our canoes are being carried away by a powerful current toward unknown shores.

"One day some of these black people with white souls came with great metal buffaloes. These creatures tore up trees in huge numbers, and in a few days the earth was completely naked. Its deep-red flesh, which our hoes never exposed, was now turned up for all to see, as though the earth was stretched out on a butcher's slab. We had only to look at this earth to know it was suffering. But our sons were amazed by this and had only admiration for it. Our giant trees were cut up like animals to be sacrificed at a feast. We felt no amazement, only sadness to see their pieces being hauled off by noisy, smoking metal mules. A great silence filled the sky where their branches used to rise. Then that silence became an agonizing absence.

"Our children came back from the big villages, where the whites had explained to them why this was for our own good. 'Look, here is what we got for the trees,' they said and showed us much larzan. They said we could use it to buy useful things for our village. They also brought with them strange food and a kind of liquid that looks like blood and makes people crazy if they drink too much of it. As time went on, our children seemed to want more and more of this drink. Plenty of grain, a forest full of fruits and game, water with fish swimming in it, enough clothes for their women and children, and a house to live in—these things were no longer good enough for them. Sooner or later, they grew bored with the things they had and wanted new things.

"In the old days, the village followed the seasons, the rising and setting of the sun, the coming of night and the dawn of a new day. But in the new time people were seized by a kind of haste. The metal donkeys continued to destroy the silence, as they destroyed our old trails. More and more ugly and noisy objects showed up, disturbing the peace of our days and even the rhythm of our breathing. Our hearts no longer beat properly; instead they sounded like drums with torn skins.

"Many years passed in this way. Finally, people began to say that the whites dominated our earth and were foreign occupiers. This made some people angry, especially the black people in the big villages. Then one day, we learned that our country had gained its freedom and that we should feel free too. Our brothers from the big village asked us to join in celebration. We put on our finest ceremonial garments, sacrificed animals, and lit great fires in preparation. The drumming, singing, and dancing filled the night. It was clear, though, that the parents were not feeling the same joy as the children. We saw no real reason to be happy—only our children seemed to understand why we should feel joy. At first they tried to explain it to us, but they gave up when they saw we didn't understand their reasons. Our own lives had become incomprehensible to us.

"A short time after this, a man of our color came to tell us we must all work to build a big road through the forest. He said that a very great chief, father and master of all our country, was going to pay us the great honor of a visit. Our children were happy about this, but they also were worried and we didn't know why. All the able male villagers and even many women worked on the road, filling holes with rocks, cutting brush and trees. We wondered if the great chief was some kind of giant, because the road we were building was five times as wide as our roads. This work lasted many days, and we were all tired. Finally, the man from the great chief complimented us and told us the chief would be very satisfied with us. The road we had carved through the forest looked like a scar that would never go away.

"For years, our children had been making wounds, sometimes small, sometimes big, with their metal creatures. Our land looked like a face torn by a leopard's claws. It was becoming unrecognizable to us. The animals were frightened away and fled toward distant areas where there was still some forest. We had to walk longer and longer to find game.

"Finally, the day we were waiting for arrived. They told us to prepare a feast, for the great chief was coming. We cleaned the village, bathed, and put on our ceremonial clothes. We waited for a long time. Finally, enormous metal insects with round eyes in front came onto the road we had built. We had already seen creatures like this before, but never so close. We knew that they were called *loto*.* The smoke and noise that came from their rear had always been unpleasant to us, but we accepted them. We accepted everything now, because our children admired all these things.

"From the first loto there came a man of our color, dressed in a strange way. His head was covered with a flat hat and his eyes were invisible behind eye hiders like those worn by some of our children. A number of other men got out of the other lotos. Some of them carried fire sticks. They were of our color and were very respectful of the great chief.

"Our children stood in rows like posts in the ground. The great chief came up to them and began to take their hands and shake them, one after the other. Though of our color, he spoke to them in the white language. All these men spoke, dressed, and walked like whites. Our village chief was waiting on his throne so that the foreign chief could come and make the customary gesture of honor that a visiting chief always performs. But our chief was told that compared to the great chief, he was much too small a chief for this. The men with fire sticks, along with our children, urged our leader to stand up, for this was the chief of many chiefs, the master of all our lands. Our chief became angry, but our children begged him not to show anger because it would make things very bad for all of us. So the great chief sat on a throne covered with leather and beautiful cloths woven by our women. He performed the traditional libation with some milk that our chief offered him. Then he began to speak: 'I have come to tell you that a new time has arrived. The people who stole our country have given it back to us. We can now hold our heads high and be proud. But our joy must not make us forget that dignity and prosperity come only through hard work. The whites have shown us how to work. Some of your own children have been taught the white men's secrets, and we will now teach these secrets all over our great and beautiful country. We will build many houses to teach them. There will be no more separation among our tribes, but instead one single, great people. And I will be its king.'

---

*Loto is the Batifon pronunciation of the French *l'auto*, an abbreviation of *l'automobile*.

"At these words, the people accompanying the great chief began to clap their hands, and our children quickly imitated them. We also clapped our hands, but we didn't understand why. This chief was strange to us. We wanted to see his eyes, but they were completely hidden behind his black eye hiders. He spoke loudly but did not invite us to share any dialogue with him, as is the ancestral custom. His words were born of pride and they seemed to lash out at people, animals, trees, and houses. Then they dissipated, fading into the distance from which these people came. Our men, women, children, elders, and youth were all silent. Even the animals and the insects were silent. The wind was still and the clouds did not move in the sky.

"The great chief continued: 'You must cultivate your land with more diligence. From now on, you will grow cotton and peanuts in great quantities, because our country needs them. We need many cowries so we can bring you well-being and prosperity, help your children, and be a strong country among other countries. With these cowries, we will build many big villages and many big roads. Your women will no longer have to fetch water, for the water will come to them. Illness will no longer take so many of your children, for we are building houses where great doctors will heal them. Our country will never again know fear, for we are obtaining the whites' own weapons: their killer sticks, their metal elephants, and many great metal birds that spit fire and death. We will become a powerful and respected people. In order to do all these things, we need your help. Our earth is fertile, so work it with great devotion. Stop wasting your time with your ridiculous little food crops. Instead, plant crops that will bring you many cowries. We will send men of great knowledge to teach you how to do this. Obey them and think about your children's future. Work hard, and your reward will be very great.'

"The great chief spoke a long time like this. So as not to be impolite, our elders listened and sometimes nodded their heads, but really they would have liked to say that these words made no sense to them. Our village already had enough water and food, even if wild game was more and more scarce. But the young people were very happy with these words. They raised their fists in the air and shouted in rhythm together. Their features were twisted, but not in anger—they seemed to believe that this great chief was more powerful than all the spirits. Finally, the lotos swallowed all the visitors and left on the road we had built. Our children ran after them for awhile, chanting in the language of the whites and shaking their fists.

"After that, every day seemed to make us feel more and more like strangers in our own land. We were like people lost in the forest—the more we searched for a path that would lead us home, the more lost we became. Each of our elders prepared to die in a puzzled, wondering silence. . . ."

At this point in his monologue, Tyemoro fell silent for a long time. In the distance, a donkey brayed. Some women passed by outside, carrying burdens on their heads. A few children were playing in the middle of the village, their cries and laughter punctuating the silence.

Tyemoro closed his eyes, as if looking directly into the memories that lived in his head, his heart, his belly. . . . On another occasion, he told me that his life was like a muddy pond churned up by animals stamping and drinking in it—a pond that no longer reflected the sky, as it had when he was young. He said he was certain that troubles as great as this must have a meaning that might or might not be revealed to him while he was still in this world. This did not sadden him any more, for he finally accepted his fate. His hands did not grow tense and clench into fists, as they did in the old days. Who can tell the storm where to blow or the river where to run? He had grown used to quelling, with resignation, the fire of resentment that sometimes flared up in his heart. For resentment may seem like a small fire when it is born, yet it is capable of becoming a destructive blaze of anger and discord.

He also said, "I am like a cow with my udder full of milk, hurrying to a stable where no child waits to milk me."

4

The Powder of
the Whites

"Things happened just as the great chief said," Tyemoro continued. "One day, people of our color came, riding metal donkeys. They were dressed like whites, and they all wore eye hiders. They told us that the great chief had sent them to bring us some very valuable gifts. They brought some sacks that were tied to their metal donkeys. We were very curious to see what these presents were. The whole village was gathered there. Our chief ordered us to be silent and sit and listen politely to these visitors, for they had important things to tell us. Then one of them spoke:

'According to the will of the great chief, the ruler of all chiefs, we must all work harder to make our fields larger and our harvests much bigger. To do this, the great chief has ordered us to give you the powder of the whites.* The earth loves this powder. When you spread it out over the earth, your harvests will be much more abundant. Try it first on some crops you have already planted, and you will see that we speak the truth.'

"Our visitors took us to a plot of land where beans were growing, to show us how to spread the powder. With a hoe they mixed it with the earth among the rows of beans. They told us to give the plants plenty of water and wait until the beans were harvested to see what the powder could do.

---

*Chemical fertilizer.

"Our chief told the visitors to thank the great chief and he promised to distribute the white powder himself among all the families. Our visitors left, saying that they would return in one or two moons.

"Every day we came to look at the beans in that parcel. After some time had passed, we could see that the plants in this parcel were more vigorous than those in the others. At harvest time, everyone was amazed because the plants that grew with the powder gave twice as many beans as the others. Our youth were even more excited because this food for the earth would make our village wealthy. They proclaimed the good news everywhere and even made a special visit to Naori, the dean of the village elders, to obtain his blessing. He was a very old man, blind and crippled. He rarely left his house anymore, joking that in this way, death would not have to go looking for him. He listened very patiently to the young people's story and finally replied:

'I rejoice with you, though not all gifts that come from the whites are good. I no longer understand what is happening in our village. My head cannot follow so many changes. But we must welcome the ones that are beneficial, and because this powder allows us to harvest twice as much, it will make our work easier. It means you will only have to plant only half as much land.'

"These words startled the young people. They shook their heads in dismay. One of them tried to explain that, on the contrary, it meant that they had to plant even more land, to make larzan. But the old man did not accept this argument. Finally, he wrapped himself in total silence, as if to courteously beg their pardon for not understanding. Already he seemed no longer of this world, and yet none of us had been more in this world than old Naori when he was younger. He could read the forest as though it was a book, and he was a true sun of the earth and the sky. Every tree was his friend; even wild beasts obeyed him and fish seemed to swim into his nets. He knew all about plants—the ones that kill, the ones that cure. He was one of the few who could spend the darkest night alone in the forest without fear of tooth, claw, or *djinns*.* When our children began cutting the trees, Naori was one of the main protesters, but then he grew silent. This was about the same time his sight began

---

*Spirits.

to fail. To him, it was like seeing the sun going down. We all knew that Naori suffered deeply at the loss of the trees. Our children had wanted to offer him some consolation by telling him about the powder of the whites and perhaps hoped to gain his pardon as well. But for him, the earth had been massacred, and he preferred to die with it. The sooner, the better.

"Naori died three days after this visit. We can all testify that his was no ordinary death. Naori was our deepest root, and with his death, we knew in our hearts that the old times were truly gone. It seemed that even the birds, the animals, and the trees were crying like orphans when he died. Yes, they were crying, in their own way. Even our youth were affected. Their usual certainty and arrogance were strangely subdued on that day. People tried to speak to each other about it but were somehow unable to. We returned Naori's body to the earth. He was curled up like a fetus, as if waiting to be born in another world. We lamented our own failure to record and preserve the contents of his memory. We knew that this death was like a blow that split apart both the tree of our tradition and our community.

"One day, the men who had brought us the powder returned. Our chief told them: 'We have not been disappointed by you. You kept your promise, and we have had seen an abundant harvest from this powder. Next season, each family will gain from using it. We have a problem among us, however. Some of us say that this powder will allow us to reduce the size of our fields and still harvest the same amount. Others, mostly the young people, say that we should use even more of it and enlarge our fields. But is it right to grow so much that we can't even store the surplus in our two-year silos? Does not such excess violate good sense? Will it not exhaust our Mother Earth, as children exhaust their mother when they don't stop nursing, even though they've had enough?'

"Our visitors replied: 'Have you already forgotten what the great chief told you? He said to cultivate greater fields to make larzan for the good of our country. You must no longer grow food only for yourselves. The great chief has also sent you sacks of cotton and peanut seeds, and he wants you to plant them in very big fields.'

"As with the powder, the sacks of seeds were distributed to each family by our chief. But this time, one of the visitors stayed behind because his masters had told him we needed his knowledge to learn how to farm in the best way. Our children were happy about the presence among us of a man with such great knowledge. He taught them many new things. Soon, metal buffaloes called *trator** began to arrive, along with other strange metal creatures that made a great deal of noise and smelled bad. For the first time, we watched as the visitor led a vicious metal dog that growled and roared as it chewed ferociously at our trees, biting them into pieces and spitting them on the ground.

"Even the tall trees finally surrendered to this metal dog, cracking and leaning until they fell crashing upon the ground. There they lay, abandoning the sky forever. These metal creatures frightened the older people and fascinated the youth. They all had to be filled with a liquid that smelled terrible and spread a nauseous odor over the entire village.

"Dozens of moons passed. We became more and more resigned to all these disruptions. Our land began to look like an animal burned in sacrifice. Every year, a little more of its covering disappeared and the sun's rays bore down upon its naked flesh. There was no more shade in which we could walk. Fire, choppers, and metal creatures had ravaged it, scattering its entrails everywhere. And the visitors were always pushing the earth to produce more cotton and peanuts. The land was covered every season with the powder of the whites, and the harvests were better and better. But then the plants began to get sick more frequently than before and were attacked by more and more insects. Then the great chief's men showed us how to kill the insects by spreading poison over the plants. Sometimes animals were killed by these poisons too, and even some people felt sick from them. One day, a clumsy man spilled this poison in the river and it killed many fish. Men came from the great chief and took away our harvests of cotton and peanuts, giving us larzan of paper and metal pieces. They no longer gave us powder, seeds, and poisons but instead exchanged these for some of the larzan they had given us. Every year, the harvest was supposed to bring us plenty of extra larzan to buy our supply of powder, seeds, and poisons. The work of keeping up the

---

*A large chainsaw.

cotton and peanuts was hard and left us no time to grow our own food. Only a few old women continued to plant beans, onions, yams, and okra in small gardens. Our children knew how to go to the big village and use our larzan to buy food for us, but soon there was not enough larzan to buy the food we needed—the powder, seeds, and poison cost so much. Other things cost a great deal, too: the talking boxes, the bracelets that measure time, and all the tools we had to buy for our new crops.

"Never had people in the village worked so hard, and never had we known such lack of food. We sweated more than ever before but had less and less to show for it, and our stomachs suffered for lack of the good things that nature had given us in plenty before.

"We tried to save by using the seeds from the mature plants, but we discovered that the seeds we gathered from them were not fertile enough. Many of us started having bad harvests. We spoke of this to the man who took what we had grown. He told us: 'These seeds are supposed to give you only one good harvest.' This amazed us because ever since the time of our earliest ancestors, we had always been able to use seeds from our plants to grow new crops. The man's news meant we had to use our larzan every year to buy new seeds.

"A new worry then came to increase our hardships even more. Our attitude toward our own land began to change. People began to think of their land as their own and became jealous of one another. Before, the earth nursed us as our common mother, but now we began to demand more and more of her to have enough larzan to buy all the powder, seeds, and poison, and many useless objects as well. Some of us became weak and others strong because some of us had more larzan than others. Now, a man standing in the middle of our fields could see nothing but cotton and peanuts all around, as far as the horizon.

"The trees became our enemies. They got in the way of the trators that sometimes came to plow our land in exchange for larzan. Sometimes these trators grew sick and stopped, paralyzed. Special doctors would come then to cure them, but a few times the creatures died and were left to rust slowly in the field.

"Some of us could not even pay for the powder, seeds, and poison. The new seeds required much water from the water spitters that had been brought by the great chief's men. Sometimes some of our wells did not have enough water. The bigger some fields became, the harder it was for some to pay for all this, and they had to borrow. Finally, the chief's men took away the land of those who could not pay their debts. For the first time, people among us were living in misery. The old people had long ago grown silent.

"One day, my cousin Sanisi came back from a hunting expedition. His face was twisted with anger and grief, but he would not speak. When we insisted, he finally told us that he could no longer hunt, not even far away, where there was still some forest left. Every time he went to the forest, he was finding more and more creatures that had been killed uselessly. He had seen whole elephant carcasses lying on the ground, rotting—killed for only their big tusks. He saw animals that had been cut, their flesh left to rot, because people wanted only their skins. Sanisi could not understand why people now saw these animals as only a way to obtain larzan. They were our fellow creatures and had always accompanied us on the road of life.

"He began to speak in a loud voice, almost shouting: 'Why are all these animals being slaughtered, why are all these trees being killed, and why is our earth burned and scarred like this? Even the sun is ashamed to shine upon us! We used to love our earth—its forest, its water, its animals. Now we have become robbers of our own houses, thieves of our own treasure. My children rebel against these thoughts of my heart. To them, I am an old creature whose time has passed. They never dare say it, but I know this is what is in their hearts and in all our children's hearts. Their hearts, their heads, their bellies, thoughts, and dreams are filled with nothing but larzan. They went to those houses in the big village to learn the language of the whites, and there they were told to reject us, to reject the ways of their ancestors. Now, they are neither black nor white. They no longer have the torch of their ancestors to light their way, and the torches the whites have offered them give more smoke than light.

'I have been a hunter since I was a child, taught by my own father. Yet now I feel much more afraid of the dead animals than the living ones. When I was young, I learned to hunt animals by trickery and speed, and sometimes I was tricked by them instead. I learned to wait patiently in ambush, and sometimes the changing mood of the wind would send my smell to them and ruin everything. I learned to be careful of their sharp eyes and the dangerous speed of their claws or horns. A hunter, the son and grandson of hunters, I knew the joy of a successful hunt just as the animals knew the joy of escaping me. There was a balance between life and death, and all things happened through God. Whether we had meat to eat or whether we had none—it was all part of the will of the Great Designer. The animals' strength became our own strength when we cooked and ate their flesh, and its good taste carried a message of life, death, and joy that cannot be said in words.

'We were dancers in the great dance of eternity, giving and receiving. Even our own flesh was not really ours, for one day it would have to be given back to the earth, to feed the plants, the trees, and other creatures, children of the earth just as we were. When we sacrificed animals, it reminded us of our place, and their blood draining on the ground was more than blood—it was a promise of rebirth. The skin, the horns, the flesh, the bones, the entrails—none of it was wasted. Now, we find these animals rotting, wasted, killed only for their teeth or their skin. This is a very bad pollution and goes against the order of things. I have even heard from someone who came from the big village that there are whites who come here to kill only for pleasure. They have very powerful fire sticks that can kill an animal far away—the animal dies a faceless death. These hunters return home and bring pictures of themselves standing with one foot on a dead animal, to show how brave and strong they are. What do these people know about a lion, an elephant, or a leopard? They see a living creature from only a long, safe distance, and then they see a bloody carcass. This is not courage; it is pride.'

"Sanisi had spoken long, emptying himself of his anger. We all listened to him in silence. His loud words were a storm whose force bowed our heads, like trees bowed in a strong wind.

"Since that day, Sanisi has joined the clan of the silent ones, all those whom the new times have made mute. His bow, arrows, and spear now hang on his wall, waiting to join him in the grave.

"It was around this time that many of our youngest children left for the big village. Some of them were given identical clothes to wear. A few of the older ones went to houses where they learned to use the killer sticks. They became guards in the big village. Other older ones learned to go every morning to a big house, where they spent all their days and left only in the evening. We were told that they were the keepers of the new times. Not far from the village, the great chief's men built a big house, where our grandchildren went at the age of initiation to learn, just as their fathers had learned in the big village, how to write words in the white man's language. Even some of our girls learned to write. We were told that children who learned to do this better than the others would become important people with great authority and much larzan. But the more our children learned, the more ignorant we felt. They never told us we were ignorant, but we could tell by the way they looked at us and treated us that they believed we were.

"Lobi, our *griot*,* refused to tell his stories in the presence of anyone who wrote words. Yet without him, none of us, old or young, would be able to tell the story of our people. He remembered where everyone came from, who was the child of whom, all the way back to the ancient times when our community was only a fetus. He was the last living memory that held the story of our past. Until now, every generation had a griot, a living memory, and every griot had taught younger people to be griots. But now Lobi could find no young person who was able to listen with a silence and respect that were deep enough."

---

*An African storyteller.

5
Drums in the Night

"Lobi was also the man who drummed in the night. His chanted words mingled with the sound of his drums so that they went deep into your ears, all the way down inside your body, to fill your belly, your chest, and your heart. We would listen to his words as we sat around a big fire. In this way, we knew that the past was present in us—that there was no break between us and the distant past of our ancestors. We all felt like beads on the same rosary, like seeds of the same plant. Lobi told us of the unity of all things and sang out every detail of our tribal tree, with all its many branches. He sang of alliances between the living and the dead. He sang of the origins of all humanity, of a time when all men and women were born of a single soul, when the earth was our house and the sky its roof. The earth was populated by many creatures who lived with us in our house, but we did not dominate them. Some of them were mobile, some were immobile; others were spirits of air and water, light and darkness. The house was in us and we were in it, and nothing could disconnect us. This is how he sang to us of the order of things.

"But now, when we consider this new order, we feel that nothing is in its place. And we have surrendered completely to it. Some of us have even tried to force our children to learn to read and write against their will, telling them that the whites are guardians of the future and that learning their ways is our salvation. To bring these children into line with the ways of white learning, some have consulted Chief Toubou.

"Toubou was the man chosen to lead us with the blessing of the Great Designer, who also gave him the keys to the harmony of our people. He watched over us like a father and a mother watch over their sleeping children. His house was set apart from the others, on an elevated mound. From there, he could see the whole village and also beyond, in the distance. He was not there to warn us of enemies from outside, but of an enemy inside—an enemy with no face, that carries no weapons. This enemy is invisible and lives in our hearts and our bellies. He plants seeds of resentment and jealousy and makes men strike women or insult other men. He turns words into poisonous snakes and puts the spirit of the jackal into people's eyes. Then our chief must chase away this bad spirit. But if he is not very careful, this spirit will multiply, and a whole village will go insane, with people wounding and killing each other.

"Toubou managed the lands and gave each family the parcel it was due. He received these commands in his sleep and no one ever went against his word. Once every 120 moons, Toubou put on clothes that could be washed clean only by the rain. With a staff in his hand, he spent a tenth of the day walking, making a circle around the sacred forest. For a certain period after this, no one was allowed to enter the forest. It was protected by his circle, by the spirits of creation, and by the will of the Great Wizard. The few who, over time, had trespassed during the forbidden period were struck by strange afflictions: madness, misfortune, or even death. During this period, the animals, medicine plants, and wild fruits of the forest became abundant. Thus each generation was assured of the renewal of our Mother Earth, whose udder became full once again for her children.

"Chief Toubou was the weaver of our community's destiny. He made sure that the warp of the sky was woven with the weft of the earth. He also repaired broken threads, kept a watch on the colors, and balanced firmness and flexibility. He dwelled mostly in silence. His eyes saw deep into the unknown and his ears received messages no one else could hear. Chosen for this role by his predecessor when he was young, he had learned to live with great frugality except during sacred feasts of marriage, circumcision, or initiation of warriors. This frugality caused him to live with a constant tension, as if he was a taut bow, ready to send its arrow from the world of humans to another world. He was allowed intimacy with a woman only every twenty-three moons. And there were many other restrictions imposed on the chief so that he would always be vigilant for his people.

"On days of great celebration, Chief Toubou always taught the people of the village, who gathered together for such occasions. He explained the great laws of the world's design: how earth and water gave rise to breath and how the sun was born of its own rays. Our children respectfully disagreed with Toubou's teaching, saying that the earth was a round ball turning in space, as the whites had taught them. These contradictions never disturbed Toubou, but they did cause him to withdraw a bit more into seclusion.

"So a group of men and women arrived at Toubou's house to seek his counsel on the problem of their children who did not want to go to the houses where they would be taught to read and write the language of the whites. After listening to them carefully, Toubou replied:

'You cannot force the lion and the antelope to marry and live together in the same corral. How then can we make a marriage between our knowledge and that of people who live in a faraway country, where the sun's warmth is ruled by the great cold? We know nothing of their world. Sometimes I think these people with white skin and strangely colored eyes and hair must not yet be mature. Perhaps they are white because they are still in a larval stage. They are not as mature as we are, but they will be when enough time has passed, when the sun darkens their skin.

'They must have lived a long time underground, for they are masters of metal, using it to make all sorts of creatures that run, swim, fly, bite, and kill. We have never seen these creatures defecate. Even the greatest and oldest trees are helpless before their power of death. The teeth of these creatures bite the ground and grab huge mouthfuls of earth, yet even a weak man can control them with small movements. The wind that comes out of their bowels poisons the air we breathe. So you see, the whites worship things that come from deep inside the earth: metal, sickening liquids, stones, and fiery winds.

'They have made an alliance with the forces that live in the darkness of the earth and have called up powerful djinns. They are not as interested in what lives upon the earth as in what lives deep inside it. They awaken these dark, inert spirits sleeping under the ground, which then rise up to torment and corrupt everything that lives through the spirit of the Great Designer. Look what has happened to our land, so naked and scarred—even the greatest of all elephant herds could not have done so much damage. The rain that falls from the sky does not stay in the earth as it did before. Now, it runs away quickly in little streams, and these streams empty into the river, where the rain leaves us forever. This is because the earth is naked, like an animal that has been skinned of its fur. Sometimes you can even see its bones. It may be dying. Perhaps it is already dead.

'As for what you should do with your children, I have nothing to tell you. Perhaps we have to accept that they are lost. Or perhaps they are not lost. Nothing happens except by the will of the Great Designer. Some people say that the blacks who are more like whites are bringing good things to our lives. I had a long talk with an important man from the big village. He came here to ask me to tell you not to resist the new ways and to tell you to send all your children to the house where they can learn to write words. This man wants you to work hard for the good of the country, so we can pay the people who teach our children. But his word is not pure, for I could see that it was mixed with threat. I told him that we already worked hard before the white men came, and that the men from the big village have taken away our land, our work, and our children. In the big village, the houses rise very high and people live in them like termites. It is a place ruled by haste. People there walk fast for no reason, which is very strange to us. I saw this when I visited my grandson in his house there. His wife goes every day to the house where the chiefs of the big village make their rules and write them on paper. Every morning, men, women, and even children look at their bracelets that measure time. Sometimes when they look at them they jump as though a spider has bitten them and rush off, saying they have no time. Why don't they listen to their own breath and the beating of their heart? They would be more peaceful if they did. The seasons, the sun, the moon, the plants, and animals—do they hurry because they have no time? Why have these people created such haste? In the big village there are so many people, and every day they have to look at too many unknown faces. Sometimes

I greeted them and received no greeting in return. Why do they swarm like insects? Their houses are often empty all day long, and these dwellings are filled with strange things: boxes that capture the cold, boxes with windows through which you can see people moving, talking, fighting, or making love. When it is hot, they have round metal dragonflies that make wind, and when night comes, they push it away with torches that have no fire and make no smoke. I have seen all these things and many others. . . .

'But why complain? And why be surprised at such things? Some of us are telling our children not to tire their arms and dirty their hands working with hoes in the fields, for they must keep their strength for becoming very good at writing words so that they will make many cowries. Some of us are proud when our children do this, and we brag about them, saying "Our son lives in the big village and comes to visit in his loto, bringing many presents. This makes him important, and respected by everyone."

'My grandson explained to me that part of the larzan from the cotton and peanuts must be given to the whites in exchange for their metal creatures and fire sticks. One day in the big village, I went to a big open space and saw a huge number of men there, all carrying fire sticks. They dressed exactly alike and walked exactly alike, in perfect rhythm to drums—but not drums like ours. Their bodies were stiff and they moved only their legs and the arm that was not holding the fire stick. They had put up a huge pole in the center of this space, and a man stood there, yelling at them. With every yell he made them change their movement together, as if their legs, arms, and heads were all attached by an invisible string. They said nothing and looked at nothing; they stared straight ahead as blind people do. There was nothing in their faces—no pain, no joy, no anger, nothing. They looked like dead men. Then they stopped and the man yelled again and they lifted their hands to their foreheads at the same time, as if some pain had struck them all. They also slapped their thighs and their fire sticks. A man blew into a shiny, very loud horn and another man pulled on a rope that raised a piece of colored cloth to the top of the pole. I understood that these men were behaving in this strange way because of their devotion to the colored cloth. I asked some people if this cloth had been left by some great ancestor. But they only smiled, and I could see they thought my question was foolish.

'That day, the great chief spoke to everyone there. He said we should all be ready to die for the colored cloth, because it was the sign of our country. I did not understand this, but I knew better than to ask anyone there to explain. I knew they would mock me. The great chief spoke for a long time, and then the people started yelling and clapping their hands in approval. The clapping sounded like a rainstorm falling on the houses. I heard some people say in secret that this chief was not the right one and that it would be better if someone else were chief. But they were careful to whisper because some people who had said this too loudly had been taken away by men with fire sticks and closed up in big houses with high walls. An old mother said her son was being held in one of these houses and she had not seen him for a long time.

'I told my grandson that I was surprised and could not understand why we should want to exchange for fire sticks the cowries gained from our land's harvest. He told me that our life depended on our ability to kill, for people in countries not far away wanted to come take away our land, our wives, our children, and our goods. He said those countries already had many fire sticks. In some places there had been great battles between tribes with fire sticks. Many men, and even women and children, were killed. He told me that in the countries of the whites, even more people had been killed by metal birds that dropped thunderbolts from the sky. I told him that it appeared to me that people in all these countries wanted the earth to give them life so they could use this life to make death. He laughed and said yes, that was more or less the way things are everywhere.

'I went for long rides with him in his loto. At first, I did not want to get into this metal creature's belly, thinking I might not be able to get back out. But then I got used to it. Sometimes we went so fast in it that my eyes could not even hold onto the things that passed by like a raging river. This was unpleasant for me, but little by little I accepted this, and many other things I could not understand.

'Now, I have no more advice to give you. I am the last bead on the great rosary of our chiefs who lived in peace with the earth. I have no disputes to resolve and no land to manage according to the will of the

Great Designer. Today, there is not a single one of our children with an untroubled heart. This is why there is no one to receive my charge and take my place. Perhaps we are like travelers lost in the forest. Perhaps this is a time that demands that we stop and reflect. Surely there must be a way out of this. It cannot be that there is no path to take us out of this forest, for that would mean that all life must die. And if all life dies, then death must die too.'"

6
Drought

"All of us remember the time," Tyemoro continued, "when the rain first refused to fall upon the fields we planted. It was not so long ago when it began. We had all done our very best. All of us, with our hoes and our animals, had worked hard to prepare our parcels of earth, doing our duty—but the sky seemed to have forgotten us. We looked and looked at it and never saw clouds. We looked at the dry, thirsty earth without being able to do anything about its suffering. Nothing is left of the land of our ancestors except our troubled memories. What we see now is like a great leper's skin, spread out as far as the eye can see.

"We had known droughts before, but we did not understand why now, every year, the wind brought so much dust, and why the sun was burning so much hotter than ever before. The first planting gave us only skinny and fruitless plants, and the same happened with the next plantings. Unlike in the old days, because of this new way of growing crops for larzan, we had no reserves of grain. Soon we had no larzan left to buy food and there was nothing to feed our children. Our livestock ranged far and wide over the dry land, kicking up clouds of dust, looking for scraps of grass or bushes to eat. We wondered then what we had done to deserve such terrible things. Perhaps we have done something to anger the spirits who stir the great cauldron of life in the invisible world.

"We were not the only ones affected. Many other tribes, many other lands and animals also suffered. Some years the troubles multiplied. The earth was split, with great cracks as far as the eye could see, and animal skeletons could be found here and there. Many trees died. Men, women, and children also died, and many of those spared by death left their land to wander the great roads, looking for some place where there was life, hoping for a miracle. Their eyes had a great darkness in them, full of sorrow. Everything was emaciated: the earth, the animals, the people, the trees. The ponds dried up and turned to baked mud, hard as old bark. We performed many rituals and sacrifices on our altars and in our sanctuaries, but nothing could stop the suffering.

"Even our master of harmony himself had no understanding of the meaning of these things. He withdrew into silence, as so many elders had, waiting to leave for the great voyage.

"One day, some whites and blacks arrived together in metal buffaloes filled with sacks of grain, which was supposed to save us from starvation. Some of it looked and tasted strange to us, but we were so hungry that it didn't matter. Our women cooked these grains with some herbs and fat and this saved us, but we still did not have enough to eat well, and we were very weak.

"Many other blacks and whites then came to help us, to teach us how to better organize our lives. Some of them dug wells and built in the rivers those big walls they call dams. Others came to tell us we had to work hard to become prosperous again. But we had not forgotten the times before when men spoke to us like this because they wanted to get cowries from us. Our sweat, our land, our water, our trees, our animals, and children have been used only for the prosperity of the big villages and their termite people.

"There were many foreigners who wanted to be generous and came from other countries to help us. We began to realize that much of their own prosperity came from what had been taken from us. Thus the parents set our houses on fire, and the children have come to try to put it out. But we know that this fire cannot be put out by foreigners coming to help. We must do it ourselves, because it is our lands and our lives that are burning. It is useless to harbor resentment and bitterness for what has been done.

"For a long time, we thought that this help from people far away would save us. But now we know that there is trouble mixed in with their generosity. Some of them always talk about their god, who they say is better than anyone else's God, and others talk to us about great ancestors in countries far away, saying that the words they have written will show us the way. Many of these people speak to us of their friendship and concern for us, but then we realize that they have secretly found some new way to milk our land dry. We know now that total depletion of our land is not far away. Immense quantities of cowries have been made from it and have been spread around, like the bloody entrails of our Mother Earth. These cowries have made a tiny number of blacks and whites very rich and have made us very poor and have destroyed our land. Now we see this, but our eyes have been too late in opening.

"The future is in our own hands. We must find a way to paddle our canoe against the current, to recover something of our life. This will demand all our strength—all the strength of the women and children who remain with us. We cannot continue living between life and death. The time has come for us to either heal ourselves or return to the Eternal."

After these words, Tyemoro seemed to want to say no more. He gave a small sigh—not a sigh of regret or pain or nostalgia, but simply an unmistakable gesture of closure. I understood that it was time for me to go. Yet his eyes also seemed to tell me to be patient, for he would have more to say later, when his well had replenished itself from its source.

I left the old man after holding hands with him briefly. I loved the way my friend would take both my hands in his and hold them silently. When he did this, I felt strength, warmth, and peace fill me. It was as if all time and all history were suddenly concentrated in this brief physical gesture evoking all that was still alive, and well.

I left his house in the twilight. Outdoors, the sounds were growing softer with the coming of night. Only the noise of a distant motorbike, like some raucous insect, troubled the silence and peace. On the horizon, a sun veiled by dust was sinking behind the great hills.

It was a mystery and a miracle to me that Tyemoro was able to tell in such detail this story of despair without despairing himself. I was profoundly moved by this tale of ruin and destruction, yet paradoxically, it left me with a feeling of strength. Certainly the situation was very bad, yet I did not feel the story was over. Somehow I felt that circumstances were not irreversible.

I knew that if I was to talk to Tyemoro again, I must leave the initiative to him and wait patiently for his summons. More than a week passed before I saw him. When we did meet, though, he made not the slightest allusion to anything he had said to me during our previous visit. It was as if he had never told me the story. I knew from experience that Tyemoro was a man whose word grew naturally in him, organic, whole, and indivisible. And when he had adopted his customary posture and gave birth to his word—or poured it out, as he put it—he never revised it or went back on it. He did not like digressions or commentaries on what he had said. In the beginning of our relationship, this disconcerted me. Why did he close our exchanges with a hermetic seal? Tyemoro never liked to chat, and even in requesting the needs of daily life, he would say nothing if a gesture would suffice. His wives, daughters, and grandchildren were perfectly adapted to his ways and attentive to his signs. They were totally devoted to him and they knew that they were all dear to his heart. With only his appearance, a quarrel in his house or among members of the village stopped instantly. Whenever a serious dispute arose, he always found a way to settle it.

All of them also knew in their hearts, without the shadow of a doubt, that Tyemoro was capable of giving his life for each of them. Sometimes his dignity and profound compassion were troubled by a brief tremor in his expression that revealed grief or pain, yet peace and well-being never ceased to sparkle in his eyes, as if even his grief were of little account. And this was only one of many other mysteries that characterized him.

My relationship with this elder had become almost like a drug to me. When we were alone together, it was as if my own personality disappeared. What had begun as an inhibition of my own expression, however, had become a blessed, liberating experience for me. Giving myself wholly to the force that emanated from him, I entered into an altered state beyond time. I often wondered about the nature of this force, feeling that it originated in the earth itself, from real and deep roots in an ancient land and culture. Perhaps this is why when I compared myself to him, I felt a certain frailty in myself.

7

A Strange Rebirth

My very first encounter with Tyemoro had been difficult. We came from such different worlds. This difference gave me a measure of the artificiality of our so-called modern life, which we wear like a straitjacket—a programming so rigid that the guidance and reference points provided by nature have become distant and foreign to us.

My head was full of theories, diagrams, and preconceived notions when I first encountered Tyemoro. Probably I was unconsciously caught up in my role, an entomologist of sorts, eager to examine a live specimen of a species I had already studied and thought I knew well. I already had fantasies of delivering eloquent and memorable lectures before large numbers of students who sat mute with admiration.

The old man seemed to see through this ruse, however, implicitly refusing to play my game or cooperate with my ambitions. He was not fazed by my knowledge of his language and culture; at times his kindly, amused smile seemed to show that he completely understood the situation. Yet this barrier had nothing to do with defense. It was simply and calmly there, neither good nor bad but a natural consequence of Tyemore's simple devotion to the truth.

Yet at the time I did not see it this way. My initial failures as a field researcher were extremely painful to me and I felt real anger at Tyemoro. I left twice with no significant material, and the second time my most adamant determination began to give way to a sense of despair. I inflicted a chronic bad mood on my wife and children, and at the university I was haunted by a sense of shame at my incompetence. But this frustration became so great that it also forced me to deepen my reflections. For the first time, I sensed I was headed for shipwreck and seriously considered abandoning this whole field of research. After all, Batifon is an obscure language spoken by a tiny minority, and informants like Tyemoro are extremely rare.

In my mental fog, images of certain scenes with him began to recur spontaneously in my mind, over and over, in precise detail. Suddenly, it occurred to me that I had never received any sort of formal refusal or rejection from him. Then it dawned on me that what the two of us lacked was an authentic, human rapport between us. Since Tyemoro had never asked anything of me, it was up to me to find the right tone or wavelength to establish this relationship. I was the one who had to get in sync with him, to tune my instrument to his, so to speak. I could do this only if I was totally honest and sincere. This old man was not some sort of gold mine of information—he was an initiate. I finally saw that we could establish some kind of rapport only if he fully trusted me, with a sense of friendship and ease. I would have to forget my old agendas. The quest for a human relationship would have to become more important to me than my quest for information and results. Realizing this, I had a kind of premonition of the calm and intimacy of our future encounters, when this old man and I would come together to serve the word that transcends mere information. I would have to collaborate with him, eluding the trap of passing time, in order to enter into the timeless realm of myths and symbols.

Burning with the fire of this transformation within me, vacillating at times between crazy elation and pessimism, I spent two years experiencing a difficult rebirth; I was pounded into shape by the obligations of daily life and hardened and glazed like pottery by the fire of time. Finally, awake at last to the true nature of my mission, I boarded a plane for the third time to visit Tyemoro, strengthened by a will that was not willful and a purpose purified of hidden agendas.

When the old man saw me this time, his face lit up with a broad, magnificent smile. He embraced me and held me against him for a long time, as if transmitting his energy to me. For the first time ever, he called me his son. Then he spoke: "So, you have returned! But now you are a true receiver, for I see the greed has gone from your heart. You have become like an empty vessel, and you are no longer a chaser of words. I am ready to give you the water I have, but you must use it for a sacred purpose. Later, you must be careful and make an effort to share these words with people in a way that fills their heart with respect, for I am speaking of the substance of our ancestors and the principles that inspired them.

"Yet you will have only my words, for the process of true initiation is very long. It puts to the test all our patience and endurance. It takes in the whole of our life, with all its suffering and joy. I can give you only the shape and sound of my words; you will have to add to them substance and spirit. You will always be a white man, and this is where you must find your own truth. You must walk your path of initiation, not mine."

Thus was born a relationship that was all the stronger for having come close to disintegrating. It is a relationship purified by the fires of truth, with roots in the depths of our hearts.

Finally, one evening, Tyemoro made a gesture for me to accompany him to his room. He lit a candle, for the night's darkness had become dense after a very brief twilight. Its fabric was palpable, without the faintest hint of moonlight. It was a "mother night," as Tyemoro called it. After assuming his usual posture, the old man resumed his narrative.

"What I am going to tell you this time does not come from my own knowledge. It requires much reflection, and your opinion as a white man will be useful to me, for my mind is sometimes like an unbalanced scale, moving up and down in search of the right measure.

"Not long ago, we began hearing much about a village called Mafi, half a day's walk from here. People told us it is a village much like ours, except that the people there have banished the evil afflicting us here: they are working to drive it away by putting back together the broken pieces of their dead community and breathing new life into it. This astonished us, for we had never before heard of a village dying and being reborn in this way. Those who had seen this village were full of praise for it, saying that life there is good and that everyone has enough to eat. They said there are children playing happily there, and that their fields are fertile. We were very curious about this prosperous village, so I decided to go there, accompanied by several members of my family and community—for I was very worried about the survival of what remained of our community.

"As soon as we saw the village in the distance, we knew that it was very alive. There were many trees standing there. Like a man dressed in ceremonial straw clothes, the village was spread out over a bare stretch of straw-covered land. As we drew closer, we were filled with a fresh breeze and pleasant odors. There were large gardens all around the houses; we were sure they must be lucky in the amount of water they had. Some men came to welcome us and lead us to the village chief, a man named Moulia. These people spoke our language.

"The chief was very friendly to our group and expressed a special respect for me, saying that the reputation of Tyemoro was known to them, that he is one whose breast holds the essential knowledge of our people. We conversed with him, exchanging news about our families and their conditions. I told Moulia the reason for our visit—that many of us had heard of their prosperity. I told him of the misery of our community and the obstacles we faced. He told me that their own community had reached such a state of hardship years before we did, and that they were almost ready to abandon the village forever were it not for a man named Ousseini. Thanks to his knowledge, courage, and stubbornness, the whole village had again found happiness and well-being. According to Moulia, Ousseini was the new wizard of their people, the man who shaped their destiny.

"This man himself arrived the next day of our visit. He was young but of noble bearing. I saw immediately that he was not like other young men, for his eyes had that light that comes from resolution, clarity, and strength tempered by great patience. Other people also came to join us, and finally there were a large number of us gathered under the big village tree as Ousseini began to speak."

8

Ousseini's Teaching

'**M**ay you all be welcome here,' he began. 'We hope for good health for yourselves and your families. I know that your visit is born of troubles—troubles we have known and still know. But we have shown that when men and women come together to find a common solution and head in the right direction with all their conviction, they can drive away this evil misfortune. They become like people in a great canoe who give their movements, their breath, and their force to a single rhythm, and the canoe glides lightly over the water in the direction they have chosen together. But these things never come easily, and we must be very vigilant if we wish to continue to rebuild our earth and our community. Pride must never be allowed to grow in our hearts and darken our minds. It is true that our fields are now fertile again. It is true that our trees are bearing more and more fruit. Our animals now have good coats of fur and walk with a vigorous step. Now, we rarely lack for grain, fruit, vegetables, meat, and milk. We feel that the time will come when our bodies and minds—especially those of our children—will know peace again and will awaken to that which is still greater and beneficial.

'It is not good for humans to live with the misery of not having enough food. Their minds become sterile and their strength declines a little more every day, which brings them closer and closer to the ghost realm. I was born of the roots of this land, just like my mother and her mother before that. As far back as our memory can reach, all the way to the farthest horizon of time, there are still these roots from which I was born.

'I am one of those who was very young when I was sent to the houses where the whites taught us to write words. My teachers told me that I was especially gifted for this. They said, "This child will go very far on the road to knowledge. He will be a great and useful example for his people and a credit to his race." I worked hard to learn what the whites taught me. My family and my clan were proud of me; I was their favorite when I was still quite young. They never allowed me to use a hoe, to plant seeds, or to work the earth as my ancestors and my family all did. When I expressed a desire to do so, they always said the same thing: "Don't dirty your hands or tire yourself! Keep all your strength for your head, so it can learn many things. You are no longer as we are; you are the doctor of our family body. You will save us by learning the secrets of the whites and bringing us their marvels. Everyone can see that the whites hold the keys to the future. You will bring these keys to us and teach us to use them. The whites are great masters of prosperity: You must learn their way of hunting and become a great hunter. You will make many cowries; you will be blessed and will bless us in return."

'I took these words as the truth and wisdom of my parents. Some older members of our family strongly disagreed with them, however, saying they were worried that our traditions were in danger. Too many of our children were turning away from them and toward this new knowledge. They told my parents that we must not forsake the teaching of our ancestors, for it is the only way to keep the human vessel strong and preserve its contents. New and foreign customs are dangerous and might cause the vessel to crack.

'I slipped through these quarrels as I might weave my way through a crowd in the marketplace. I was never discouraged by any obstacles that appeared in my path; I passed through all the white people's houses of initiation and went to live in the big village. I was now far from the bones of my ancestors, far from my home and its dry and thirsty earth. Every day, I went to the greatest house of initiation in the big village. I had never seen so many other students. Most were boys, but there were also many girls.

'One day the great chief came to speak to us. There were great preparations for his visit, with ceremonial dances, drums, and chants. We dressed in festive clothes to honor the supreme leader and were required to learn a special hymn glorifying our country and praising the great chief as our savior, next to God. Some students said that God was nothing but a children's tale whose purpose was to put our reason to sleep. They said we should rely only on our reason and forget about God. Sometimes this led to intense quarrels, for other students were outraged by these words.

'The supreme chief arrived and told us that we were the foundation and the framework of the house of our people. Thanks to us, he said, our country would have a place of great honor among other countries. We responded to his speech with great shouts of approval as we waved cloths with the colors of our country. Each of us vowed to become a valiant warrior of the new times and use our knowledge to make many cowries.

'My own decision was to study farming as the whites did it. I thought that with this knowledge, I could help my country to prosper and never to know hunger again. I learned many things about ways to make the land produce more and more. I believed that the whites could show us how to achieve great abundance. Other students studied the art of management, and some studied how to win wars or how rules are established between different countries. Still others learned how to judge those accused of crimes, including murder, and how to punish them, and some studied how to defend these same accused people and keep them from being punished.'"

Tyemoro now hesitated in his story. He had clearly encountered some obstacle that made it difficult to continue his account of Ousseini's tale. Until then, his narrative had flowed smoothly; I knew this pause was not due to some failure of his prodigious verbal memory. After a brief silence broken by the odd noises of animals in the distance, I realized that the night had grown very dark. Then he again took up his narrative.

"Ousseini then told us this: 'My abilities led me to be selected as one of the few sent to study in the mother country of the whites. I walked into the belly of a giant metal bird, and there were many other people inside with me. The bird lifted us high into the sky and finally put us down in the land of the whites. As soon as I arrived, I knew that the wonders we have heard about these people are really true. Their big villages are amazing, with tall houses in long rows—so tall that they seem to be holding up the sky. There are great numbers of lotos and many people walking in all directions, so that everything is busy and agitated all the time, like a giant ant bed. There are very wide, straight roads between the rows of houses, and some of them go so far you cannot find their end.

'In the beginning, everything in this place was amazing to me. Everything here seemed very different from where I was born. In these great villages, people get used to never seeing fields, forests, or animals. But some of them keep dogs and cats, which they caress with affection. I, too, finally got used to never seeing stars in the sky, and not knowing where the moon was. My house was a small place that looked like a cabinet, built and carefully polished by a carpenter. Everything was full of both light and shining objects. I was given many of them to use for making heat, light, wind, and even words and music. The water I used flowed out of many metal intestines, from where I did not know.

'Every day, I went to the great house where our initiators taught us. There were many of them, both men and women. In these great villages there are so many people that they all dissolve together. I, too, dissolved into this swarm, and I began to feel very lonely. To fight this, I made friends with people of my own color, and also with some whites. Friendship helped us to feel a little happier.

'In this great village, all children go to the houses of initiation. Adults who have been initiated go out every day to make larzan. When people get old and weak, they are sent to houses to die. Those who are simple-minded and insane are sent to different houses, where healers try to restore their minds. Those who commit crimes are closed up for many years in special houses. The whole world in the great village is filled with many, many doors with locks and keys. All people are sifted and filtered according to their condition, as grains of wheat are sifted.

'The haste in this great village is so great that it caused me to lose my way of looking, my way of hearing, and my sense of smell and taste. As soon as I woke in the morning, everything seemed to be telling me to hurry. I learned to be in a hurry almost all the time, and my sleep became troubled by a host of little demons always agitating my thoughts. People complained about not having enough time, and sometimes they were overcome by extreme fatigue. I consoled myself with the thought that all this was necessary for my initiation.

'To complete my initiation, my teachers sent me for a long time to work and study with their farmers. I was glad to leave the big village and rediscover trees, grass, wind, and stars, and feel the sunshine, the rain, and the cold. The cold was great, and I suffered from it at first. Sometimes it made my whole body tremble. But the whites who welcomed me there were very kind and gave me thick clothes to protect me. I was happy among them. The fields they farmed were so huge that they stretched from one horizon to another. They had great numbers of special trators: some to turn over the earth with claws, others to plant seeds, and still others to harvest the crops. This all amazed me, for it seemed that the metal itself was intelligent and devoted to serving humans. The chief farmer taught me how to use all of these metal creatures. But I often felt very tired, for we worked in the fields from the morning on, and sometimes even at night, when the eyes of the trators lit up the fields so we could complete any important work that we had not been able to finish during the day. The whites do not think as we do about the night. For them, there are no spirits or djinns in the paths, fields, forests, or anywhere else on earth.

'My hosts also had many real, live cows made of flesh and blood. Every morning and every evening they were milked by metal creatures made especially for this purpose. The milk flowed into intestines made of a material unknown in our land, and these emptied it into huge barrels. So much milk poured into these barrels every day that a tall man could drown in it. The cows were chained to posts, and a great deal of food was given to them so that they would give abundant milk.

'In other places I saw chickens in cages so huge that our entire village would fit in one of them. There were so many chickens that their clucking was like a roar forcing us to shout in order to be heard. Every day, huge numbers of eggs rolled from the birds' bodies and were sent to great baskets. The chickens' feces made small mountains.

'These farmers, helped by their metal servants, had a duty to produce everything in great quantities: milk, meat, grains, and many vegetables. This abundance was needed to feed the huge numbers of people living in the great villages. They also had to feed their own livestock large quantities of food, so that they would be fat and plentiful.

'I learned that much of the food given to these animals came from farmers in our country and other countries like ours. The white farmers calculated that they would pay fewer cowries in this way. I also learned that some whites strongly criticized this and blamed these farmers for bringing devastation to our lands; they believed that exchanging our grains for larzan cheated our farmers.

'After a number of moons with these farmers, I began to feel bored. There were not many of us working there because the metal creatures could do the work of twenty or more men. Most of the farmers from land nearby had left to live in the great village and work in houses where they make lotos and other metal creatures. My host had two children, a boy and a girl, but they were always gone, at the houses of initiation in the great village, and very rarely came back to the farm. All those giant planted fields, with woods in the distance and so few people living there, made me sad. There were no festivals and few people to talk to, and there were never any community meetings, as we have. My hosts complained that they had too much work to do and were not receiving enough larzan for their crops. I learned that there were many farmers who went into the streets to angrily shout their dissatisfaction. Some of them even complained about the abundance of their harvests: they said this meant they would receive less larzan for their crops. During one of these gatherings of angry farmers, I even saw them throw mountains of food into the streets to rot.

'This is how I began to learn that the world is divided into rich and poor. Everywhere in the world, the people who have the most metal creatures and fire sticks, who know how to write words well and understand how larzan works, are able to take things away from people who don't have or can't do these things.

'I learned that for every group of five human beings on this earth, there is one person who takes four of the five fruits from the common basket while the other four people must divide up the one fruit that is left. This is surely against the true order of things, for the Great Designer wants all creatures to have their due. Today, men have created an order that is out of balance; everything is controlled by the richest and most powerful.

'I believed in the goodness and generosity of humans. I told myself that if I could learn the secrets of the whites, I would be able to use this knowledge to help my own people. But what I was offered as a reward for learning these secrets well was an invitation to become one of the small number of people that takes more than its share from the basket. I saw that among these people, caring about others is rare. They are ruled by the desire to possess more and more, and I knew that if I accepted their invitation, I would become filled with greed just as they are.

'The whites have created powerful tools for learning what is happening to people everywhere. From their talking boxes you can hear about things that are happening right now, but very far away. In the window boxes you can see what is happening in other places, even in the most distant countries. The written word also covers the earth. All of these tools tell constantly of the suffering of multitudes of human beings everywhere.

'All over the earth the land has been divided into parcels and a very few people get most of the best ones, leaving the smallest and poorest parcels for the rest. These few people want to be honored like princes; many of them want to be honored for ruling some part of the world. Those who succeed say: "Look how great I am: I hold this piece of the world in my power. Everyone listens with respect to my words. Big crowds come to venerate me. They put their hopes and their lives in my hands!" Others say, "Behold my great knowledge, for I am a guiding light! Are you not dazzled by my light?" Still others exchange images of themselves for cowries. In their secret of hearts they think: "What a magnificent creature I am! What a credit to the work of nature! I will make many cowries by pleasing and amusing people, by making them forget their boredom for a while, and they will worship me. Their boredom is so great that they will use my image as a cure for all their troubles. They will give me many cowries, and I will be very important."

'There are also those who hold their heads high, and say: "We are a great and powerful people, for we have metal creatures of terrible destruction. These creatures lie waiting in secret caves, wrapped in silence. Death is their glory, and they await only our command to kill an entire country of people and its land, trees, and animals."

'Many women there are proud of the shape of their bodies, their faces, their breasts, and their buttocks. They say: "I am the dream of every man and the object of jealousy of other women. I am an antelope, which offers itself to all hunters, yet none can capture me. I haunt men's dreams, and they will give much larzan to see me and hear my voice."

'There are many masters of words there. They fill the silence with their words, admiring the sound of their own voices and saying: "How true and just are my words! I have been gifted with great intelligence, which many people admire. This gives me great authority and many disciples."

'Later, I told my teachers that I was astonished to see these farmers with their metal creatures creating so much abundance that much of it is wasted or left to rot. They answered that this is the way of the wise designers of the world, that it is healthy for everyone when people compete with each other to make more larzan.

'But there are also many whites who are against this, and they say it is wrong for goods not to be shared and for people constantly to be filled with greed for more. These whites say: "All this abundance is ruining the poor people of the world, and it does not really make us happy either. Just look how sad most of us are, even with all our lotos and other metal creatures to serve us. All we do is speak more and more empty words, look for more amusement, desperate to seize the magic fruits of time, which is rushing past us like a roaring river. Our great shops have so many goods in them that even our desire for them turns to boredom. We have forgotten the patience taught by the seasons of the year, which slowly ripen fruit. We have lost the ability to enjoy life and we no longer know how to celebrate and be happy. All our desires are stirred into a confused mixture of haste by a mysterious drum beating so fast that our days rush by with drunken speed. This invisible drum beats with and against our every step. We no longer know how to relax; we know nothing of the great, slow rhythms of the sky, with its changing stars and seasons. But in our hearts, we remember these rhythms, and their loss fills us with sadness and longing."

'So I began to see that from this great abundance is born a misery that has never before been known, the misery of those who are overwhelmed by the materials people have torn from the depths of the earth. They are always saying: "Give us more larzan so that we can have more metal creatures and goods; give us more entertainers and entertainment so that we can fully enjoy our time."

'But creation cannot allow a greed without end. How can we make greed into a principle? Men and women keep feeding the fires of all these desires, keeping them burning, and they are never satisfied. They constantly repeat: "You must always have more!" And they throw all their dreams into this fire, never waiting for any miracle or surprise from life, living a life that pretends to be ordered by reason alone. "Never stop making bigger and bigger profits," they say, "your desires should never have limits, for tomorrow we will create even more amazing creatures to serve you!"

'So these men and their metal creatures maintain an abundance that is supposed to be enjoyed by others. Yet people are not satisfied, for this entire order is built on illusions and lies.

'When I began to really understand this, I felt great pain, as do many whites who understand it. We know we must refuse this order, which has no place for fairness to others, which cares nothing about the ruin and destruction it is creating.

'In the depths of my heart, I asked myself how I could honestly contribute to the well-being of my people with this knowledge I had gone to the great village to gain. My people did not have enough cowries to acquire enough metal creatures or enough powder to make their plants grow bigger or enough poison to protect the plants against disease and insects. A great suffering filled me, for I felt that I would have to return home like a hunter who returns empty-handed. I knew that my voyage of initiation had cost my country much larzan, and that this really came from the sweat of my own people. I realized that I could return and make much larzan in the great houses where people write the words that are used to rule the country. But I knew that I would never help my people free themselves from hunger and misery by shutting myself in these houses to write words.

'Too much time has passed, and during this time we have believed that learning the secrets of the whites and becoming like them would make us prosperous. But now we know that their prosperity comes from the things they take from others in distant countries and from their destruction of the earth. Many whites now realize this. These whites are not people of arrogance; they suffer with us and reject this order.

'One important thing I learned from my initiation is that there is great virtue in white people. They have taught us how to understand more about the world and even about our own traditions. Many of their secrets could be very useful to us. But I also saw that we must never imitate them in all things, for they have lost the sense of the sacred. They have made many metal creatures that kill, and they have sold these to people in all countries. They have filled the skies with poisonous smoke and thrown into the rivers poisons that kill the fish. The earth, which the Great Designer gave us to be our nourishing mother, has been violated by these people filled with greed for larzan. Many animals and plants that were sacred wonders of creation have now disappeared forever.'

9
The Four Pillars of Life

'Thus the four pillars of life—earth, water, light, and breath—have now become sick. Humans are also sick, because their hearts have become hard. Every day, the world is broken a little more, for larzan has become more important than human beings, than creation, than the Creator.

'This is what I learned from my initiation. But how could I use it to find a way to honor the trust of my people, who, ever since my childhood, have placed in me their hopes for a better future?

'Another thing I have learned is that all over the world, people are abandoning the earth and moving to the many big villages. They say: "The earth does not feed us, and it is boring to live in the bush." In the big villages, their numbers are so great that they cannot even recognize each other. This situation creates anger among them, for a small number of them live in big and beautiful houses, with shining metal creatures, while others have nothing—they are hungry and have no work. Many of them do not even have houses to live in. All manner of evils fall upon these people—hunger, disease, crime. They become beggars and even murderers. The great villages are full of fear, and many guards are there to beat people and shut them away.

'Many children in these big villages live on garbage. Some people who have studied these problems say there are too many human beings, that the earth has become like a cow with too many calves to nurse; that her udder is drying up so there is not enough milk for everyone. They say: "Humans must stop having so many children. They must learn to limit the number that are born." I think this is probably true, but I also know that the amount of waste in the world is very great: Many humans could live on the food that rich people throw away every day. I knew that many hungry children could live on the food that is used to feed animals. But people have now become so used to this situation that it seems normal to them. Except for a few, they do not denounce this order of things, for their hearts and heads are asleep.

'In these big villages there are many old people who have only a dog or cat for company. Their friends and sometimes even their own children forget them. When they become too weak, they are sent to houses to die. And when they breathe their last, they are quickly put into the ground and are covered forever by the great veil of time. As for the children, their peace of mind is tormented by adults telling them that they must work hard now so that when they grow up, they will make more cowries than their neighbors.

'We must not follow this road. There is no kindness in it. The animals the Great Designer gives us for companions and for food are also suffering terribly. In huge numbers they are crowded together in great cages to give their flesh, their milk, their eggs, and the fur of their puny young. Sometimes people who want to study the workings of their bodies or find new medicines cut them up while they are still alive. Other people kill animals for amusement, sometimes in front of great crowds, who scream out their approval for the brave man who kills the beasts. They say: "See how beautiful this man's movements are as he kills and tortures the bull! Man triumphs over animal!" But when the furious bull catches the man and knocks him to the ground, people run to save the man, for he must win and must not face the same fate as the animal. So it is that some people torture animals, while others adore them so much that it twists their nature. These are only a few of the strange things I have seen—they are not worthy of being imitated by us.

'During many long days and nights, I reflected. Sometimes I felt a desire to stay in the country of the whites like other people of my color—some of them even marry white women and live in their country with them.

'I knew that our country has been lacking rain badly for many years. I knew that ours was among the poorest of countries and it was being struck harder and harder by poverty and misery. Some of these poor countries, bewitched by their leaders, begin fighting and killing those in other countries. Their great chiefs, interested only in their own power and welfare, care nothing about the welfare of the people they rule. They steal their own country's goods and sell them to white merchants. They have bled their earth and their people and opened the doors to the torments of hell.

'As for those of us who have learned how to write words, many of us are also traitors to our people. Because our people do not know our secrets, they never realize that many of us use our knowledge to serve the world's most powerful villains and thieves. Our people have acclaimed us as initiates and guides who would show them the ways of salvation—and instead we have led them into a desert where they can no longer even feed themselves. Misery now strikes them in such great numbers that some of them take to the roads wearing rags, hoping to find someplace with a little life. We now have shepherds without sheep, farmers without fields, and men, women, and their children wandering in packs like starving hyenas. This fate is not worthy of humans. We must find remedies for these evils.

'I deepened my reflections on all these things, but my pain was great, for I did not know what to do. One day, I met a white man at a meeting of farmers. This man said: "The earth has never betrayed us. It is we who have betrayed the earth."

'These words startled me and stabbed me like a thorn. Everything this farmer said was different from what I had heard before. He spoke of the earth as a living being. He said: "In these times of great troubles, when people are threatened everywhere, only the earth can deliver them from their misery. First, we must recognize that we have inflicted terrible suffering on our Mother Earth. We have even forgotten that she is our mother. Our vanity has led us to believe that we can control the future with reason alone. The fruits of our labor have not been spread fairly among people. Most of it is seized by a small number. We can now travel all the way to the moon, but the moon offers no solution to the torments that afflict us every day. We are constantly impressed by our own wizardry, but we do not know where this wizardry is taking us. More money, more speed, more metal creatures, more food production are not giving us happiness. Our weapons of destruction, which keep multiplying, do not make us feel any safer or less anxious. Each of us carries a secret burden of solitude like a heavy, burning stone. . . ."

'I learned that this man did not farm the earth like others did. After the meeting, I approached him and told him I would like to learn more about his ways. He invited me to visit his farm for several days.

'There, I saw plants and animals in good health and all creatures, including human beings, living together in harmony. There was a very different atmosphere and spirit at this farm—it was totally unlike any of the others I had seen. His family was peaceful and happy to be there. The breezes that carried smells from the fields and forest seemed to carry peace as well. There was hard work going on, but it seemed much simpler and easier.

'Later, I returned and spent a number of moons working on this farm, for I wished to be initiated by this man. After this initiation, I knew I finally had the knowledge to make myself useful to my people and to our land. I felt I could help them build a new life. I knew that everything must begin with the living earth, for she was there at the beginning of us all.

'I returned to my village, where they held a great feast in honor of my homecoming. I was the pride and joy of my family, the son who possessed great knowledge. They told me: "Now you will go live in the big village and become one of the most important people there, with many others working under you. If you do well, you may be noticed by the great chief himself, who may invite you to work for him. You may even become the great chief yourself someday! You will live in a house that is always pleasantly cool. You will have many metal insects and many servants."

'I waited until the day after this feast to announce to them that I was going to live in our village and farm the land. This announcement caused great shock and agitation. My family believed that I had lost my mind. Some of them said I had been bewitched by sorcerers. The women broke down in tears. I was cursed for wasting all the sacrifices they had made in the early years of my initiation. All of this made me very sad, but I was strong in my resolve and nothing could shake it. It was my own truth—a precious thing that lived inside me, weightless and full of light, yet also so firm that nothing could break it. During this great storm of blame falling upon me, I remained silent. From my last initiator I had learned to love the earth—but I had no idea how difficult it would be to share this gift.

'Many days passed. My sleep was filled with strange dreams, sometimes dreams of great beauty. Being at home among my people was like living in a fire. But instead of destroying my resolution, this fire made it even stronger. After consulting my family, Chief Moulia himself, along with Abinissi, the master of lands, refused to give me any parcels of land to farm. Even the neighboring villages refused to give me land.

'I had two choices: to leave, or to try to farm those strange lands shunned by everyone because they were held to be full of wicked spirits who were hostile to humans. These lands were mostly barren, but in the middle of them was a swamp. People feared this place especially, for it was considered to be full of evil djinns that had corrupted the water. I was afraid of these places, just like others of my village. Often I would look at these lands from a distance, without daring to go there.

'One night, driven by a strange will even stronger than my own, I decided to go. There was a small moon, but that night was very dark. I felt danger everywhere—in the trees, in the rocks, in the cries of the animals. My heart was beating like a drum and my face was covered with sweat. But nothing could make me turn back.

'When I came all the way to the center of the land of evil spirits, I began to speak softly: "O spirits of this place, please see that my heart has no bad intentions. I respect you all and beg you not to harm me. I am like a plant with no roots. My desire for land is very great, and everywhere it is refused me. Like you, I have been cursed by the people here. I refuse to go live in the big village, as they want me to do. I know there is nothing for me there. It is the earth that calls me. Will you allow me to make a home in your lands, if I promise to respect them?"

'A great silence followed. The moon, reflected in the swamp waters, was covered by passing clouds, then it emerged, then it was covered, and it emerged again. This happened many times. Hours passed. Finally, as the light of dawn appeared in the sky, all fear left me. A great calm enveloped me, like a soft, warm robe. I was lying on my back, completely empty of any fearful thoughts of serpents, dangerous animals, or evil spirits. The star-filled sky above me was immense. From the earth I sensed a kind of warm breath rising, and from the sky, a cooler breath descending. The two breaths joined as one. I felt them moving through my own body. I felt like a seed that has just begun to sprout. A pleasant heaviness filled me, and I fell asleep.

'The day was well advanced when I awoke. Looking all around me, I saw nothing but arid brush and a few trees. But these trees and plants had a virgin quality, for it had been a long time since any humans had been here. At times, I could barely hear a faint sound of the village in the distance, carried on a light breeze.

'When the villagers saw me return, they reacted with astonishment and mistrust. They said the whites had removed the fear of evil from me and this was proof that I had abandoned their ancestral ways. They claimed my heart was full of insolence. It is true that most whites do not believe in the power of evil spirits. They ridicule our fears of such places as the shunned lands. Yet there was no insolence in my heart. I felt and expressed no pride and no sense of victory about spending the night alone there. It is true that I have learned many things from the whites—but I also know there are mysteries that escape their understanding. They cannot admit the power of what is invisible, of what cannot be demonstrated to all. Some of them say: "Reason is the key to understanding everything. Some things may escape reason today, but they will not escape it tomorrow." Other whites say, as we do, that there are visible worlds and invisible worlds and our reason is too weak to explain all. Everywhere in the world there are vain and arrogant people who are proud of their knowledge, but there are also humble people who accept their ignorance.

'After hearing about my transgression, some people refused to speak to me or even be near me for fear of being polluted by my sin. I revisited the lands of the spirits many times, day and night. All fear had left me, for I knew what I wanted to do was right. I knew that my intention was beneficial to myself and to all life; it could not be a mistake.

'It was said that people had died for transgressing the limits of these forbidden lands, and many expected me to die very soon. This was a torment to my parents. But time passed, and I was still alive. After a long period of patient waiting, and getting to know these lands, I finally felt that I could begin to work the earth. I felt a harmony between my own thoughts and everything that was happening in these feared lands, where the wind blows with such strange sounds. I felt that this land had adopted me as an abandoned orphan is adopted. I had no desire to be its master. This earth was beginning to feel like a great mother to me, a mother who included the sun, the moon, and the stars scattered like seeds in the vast field of the night sky. I sensed invisible currents flowing everywhere around me. The earth was also like a woman, a bride whose wedding is never finished. The sun offers her the gift of heat and light and the moon and stars seem to perform an endless dance in her honor. And when she is gifted with water from the sky, she reveals miracles.

'At this time, a story that I learned from my white initiators returned to me, and now I wish to tell it to you. It may be useful to you as part of the initiation I am offering you. You should not see this story as something that goes against our own knowledge, but as something that strengthens and completes it, just as our knowledge completes other ways of knowing. The vision of each people should be a light for all humans, for we are all really one great family.

'It is said that the earth is the daughter of the sun, a fragment of the sun. Some say the sun is a sort of huge furnace, born from the original furnace of creation. This is why we receive heat and light from it. When the night hides it from us, we lose its heat and light quickly. The earth is now very far from the original mother furnace, but she has kept the memory of the sun alive in her for a long time. At first, she was like an ember of fire, thrown into the sky. But the principle of cold began to take hold of her in her dance and invade her little by little. The principle of cold noticed that the earth was nothing but a skeleton made of barren rock. But the sun continued to bathe her in its light and heat from afar. Without it, the earth would be lost in the great darkness, the mother of all nights.

'Little by little, the Sun and the principle of cold finally came to an agreement and found a balance between them. This balance created warmth, and this warmth was good for the birth of living creatures. It also caused water to come into the sky in the form of clouds, and then begin to fall upon the bare rocks. The water flowed everywhere, into every hole, every crack, and into the great valleys between mountains. The breath then appeared in the form of great storms, angry winds between sky and earth, between cold and heat, between light and darkness. So earth, water, light, and breath were all there, along with heat and cold, and they were always struggling with each other. Water attempted to drown fire, but the fire turned the water into steam. The cold froze the water into ice, but then became captured inside it. The breath rushed everywhere around the earth, for it could not be captured—it could stay or leave according to its will. Days and nights followed each other, moons of moons, and years of years followed each other, in a dance without end.'

10
A New Order of Things

'Then, a new order of things arrived. Until that time, all of four pillars had worked together to create the foundations of life, but life itself had not yet appeared. The four pillars looked upon the naked, sterile rock wrapped in silence. They could not say whether the rock had captured the silence or the silence had captured the rock. The two, earth and silence, were as different as can be: one subtle and invisible, the other heavy, visible, dense, and hard. They were joined so deeply to each other that they seemed inseparable.

'Then the water said: "I will penetrate the rock, and when I am inside it, I will invite the cold to come in and try to capture me." In this way, the water and the cold broke the rock into many pieces, big and small.

'Then light, water, and breath joined with the warmth and the tiniest pieces of rock to create the first plants. They were almost invisible, but they began to spread wherever there was not water. Inside the water, other plants were born, and many, many other creatures as well. For some creatures, water is the most vital element. For others, it is earth and water together. For still others, earth is the vital element. And winged creatures live mostly in the air. But all of these creatures need breath, just as they need light and warmth.

'Time is divided into day and night. The seasons give it a great, slow rhythm, and cold, heat, dry, and wet give it moods. This is how life is ordered everywhere. Now on the rock struggles began between different creatures; some of them ate others, but each tried to preserve its own life. Today, some creatures eat only meat, others only plants, and still others eat both. In order not to disappear from the earth, the small and weak creatures either learn to move very swiftly or multiply in great numbers. Strength, cunning, and disguise are often used. The claw, the tooth, the horn, and the beak were born at this time and served both life and death. Life multiplied into forms beyond number, from those that were very large to those so small they were invisible to the eye. But these forms all worked together and organized to form a great order of life—an order that we only partly understand.

'From the very beginning, the principle of life was faced with the principle of death. Everything that has a beginning must have an end; the generations of plants and animals reproduced themselves in chains in which life was always dying, yet never dying. These great chains were founded on seeds created by the male and female principles. Everything died only to be reborn. There can be no life without death, and there can be no death without life.

'After a very long time, the new order became fully established. It was a pact between the earth, the plants, and the animals—all members of one great inseparable family, all links in the same great chain: no animals without plants, no plants without earth. All three needed water, light, breath, and warmth.

'In these very ancient times, there were no humans on earth. Perhaps they already existed in the thoughts or dreams of the Great Designer. Perhaps they existed in the form of tiny seeds dispersed throughout creation but were not yet able to germinate. Every tribe of humans has its own story of how we began. Some say that we all came from one couple, the parents of us all. Others say that the Great Designer made us from the jackal or the rabbit. Still others say that we were born from the shock of the first great wars of the gods. The whites say that our mother was the ancient ocean that covered the earth and in which multiplied germs invisible to the eye, germs that were the origin of all living creatures. They also say that the great ancestor of humans is the ape. Others say that the first couple, the mother and father of us all, lived in happiness at first but then disobeyed the commands of the Great Designer.

'The first humans were gifted with intelligence and feeling. They carried within them a seed of intention, and this made them able to act beyond the laws that other animals obeyed. Their intelligence helped them to go beyond the laws of survival: the constant search for food, water, and mating. They were able to hold memories of things that were gone and to imagine the future. They experienced laughter and grief. Humans were the first creatures to form images of the Great Designer and build sanctuaries to him, or to invisible spirits of animals, trees, rocks, and even the sun, the moon, the stars, and everything else hidden behind the great veil of the heavens.

'Humans were the first to build tombs to their ancestors. They were the first creatures to try to protect their corpses from the ravages of animals and decay. They held great celebrations on many occasions, praying to the spirits for protection. They feared evil spirits and had many ways of keeping them away. They stood up straight, using only their legs to walk and leaving their hands free to make knives, spears, bows and arrows, and many other tools for their welfare and survival. Without this intelligence, humans would have only the slightest means of defending themselves against stronger creatures.

'Some whites say that the only truth is the laws established by nature, and they do not believe in the Great Designer. They say that dead matter comes alive by its own laws, and that this happened by accident. They think that death is the final end; there is nothing beyond it. But most whites say, as we do, that the Great Designer is the only source and creator of all things. Some of them say that the skies are inhabited by great spirits close to the Great Designer.

'All of these beliefs have led to many quarrels among humans. Each group claims to have the only truth about all these things. Some of these groups venerate great ancestors who were their initiators in these things, people who have written their words in many books. Every one of these great ancestors has said that people should not inflict suffering on each other, but these groups build houses to the Great Designer— houses that are open only to those who believe as they do and are closed to those who feel differently. Every one of these groups says it is the protector and guide of human spirits, that it can lead the way to obtaining life without end and escaping punishment without end. Each makes many rules that must be followed in order to obtain reward and escape punishment. These rules are not the same in all groups. Each of them claims that the word of their messenger sent from God is the only true message of God. They each have different songs, celebrations, special clothes, and customs to express their devotion.

'So human beings appeared among the animals and plants of the earth, and they brought with them troubles. In the beginning they lived only on the flesh of animals and fish and on the fruits and vegetables they gathered. These were abundant, for there were many great forests then. Their intelligence also enabled them to discover which plants were useful, which could cure illnesses. Human beings brought words into creation, for nothing on earth had names until they arrived. They brought celebration, but they also brought anxiety and the fear of visible and invisible things, and many other troubles as well.

'In the beginning they did not feel that they were distant from other creatures, but their intelligence gave them the power of making and keeping fire, which gave them mastery over light and heat, unlike all other creatures. They used it to push aside the shadows of night and frighten away animals that threatened them. They began to use it to cook their food and transform things from the earth into useful tools. This gave them much prosperity. But there were many mysteries their intelligence could not understand. They knew that they would die some day, and they were afraid. Some people say that from this fear was born all thoughts of other worlds beyond visible life. They say that this fear created thoughts and images of immortal and powerful gods and spirits. Humans worshipped these gods in order to obtain protection in this life and in the life beyond death. Whatever the truth may be, all these beliefs have become so numerous and confused that the earth now knows troubles that did not exist before humans arrived.

'In these times, people still wandered often, in search of food. They discovered more and more plants and animals useful to them. They began to gather seeds from wild plants and saw that they could make good food from them. But this life of wandering in search of food had its difficulties. Other animals sometimes ate all the available food, and there were bad seasons where there was not enough to eat. Their intelligence revealed to them that they could plant these seeds in the ground near their homes. This enabled them to watch over the plants and take care of them. They also learned to tame certain wild animals so that these beasts could give them their milk, meat, skins, strength, speed, and many other advantages. Thus the shepherd and the farmer were born.

'Much later, humans began to build villages, and then big villages. They found materials in nature to build these and learned to cook clay to make pots, walls, and roofs. Later, they built huge canoes to cross the seas and meet humans in distant countries so they could exchange knowledge and goods. These canoes were also used to carry many men to other peoples to destroy and rob their villages.

'In those days, humans everywhere had great respect for the earth; they celebrated her fertility, praying for the skies to bless her and offering prayers and sacrifices. Farmers and shepherds multiplied. They became more and more skillful at taking care of plants and animals, growing them in abundance so that there was plenty of food for the people who lived in the big villages.

'Some tribes learned to write words on stones, animal skins, wood, leaves, cloths, and other materials. These people could thus share their thoughts and feelings far beyond where the voice and the ear can reach. They used these things to spread their teachings over many lands, and rule over other peoples.

'Little by little, humans began to be dissatisfied with merely enjoying the gifts of nature and the earth. They wanted to dominate and possess her as well. They desired to conquer people who lived in distant lands, and they learned to ride animals that took them very fast and very far so that they were able to vanquish many other peoples in battle. They made better and better weapons of all sorts, and they gathered themselves into huge groups of warriors who killed people in greater numbers than ever before. This brought a new kind of suffering to the earth. People stopped saying, "We are the children of creation" and began saying, "We are the owners of creation." Some of them proudly announced: "The Great Designer made this world for us. He has commanded us to use this world as we please and dominate all other creatures. We are princes, sons of the Great Lord."

'Now people in many places began to say among themselves: "How great we humans are! We have intelligence and we walk upright on two legs. Our feet may be forced to walk the earth, but our heads are spheres, like heavenly bodies." They claimed to have knowledge that enabled them to separate good from evil, beautiful from ugly, strong from weak, and truth from falsehood. Sometimes one group's good was seen as evil by another group, one group's ugly another group saw as beautiful, truth for one group was falsehood for another. This caused many troubles. Some groups conquered others and made servants or slaves of them. Sometimes these slave masters would use their intelligence to try to understand great ideas, such as the meaning of freedom. They tried to understand life and death, harmony and disharmony, pleasure and pain. They often spoke of the good and the beautiful—but they forgot the terrible conditions they inflicted on their slaves and others. They created a new way of worship in which humans began to celebrate and glorify humanity, even saying that the Great Designer must resemble a human. Some of these people even claimed to be the Great Designer himself in human form.

'Thus the gift of intelligence, given by the Great Designer to humans for the purpose of understanding more, loving more, and bringing greater compassion into the world, was instead serving arrogance, violence, and injustice. People began to inflict more and more suffering on each other and on other creatures. Powerful small clans seized vast lands for themselves alone after killing or chasing away the people who had lived there before they arrived. Humans began to forget how they used to live and exalted force and power above all principles. Instead of respecting kindness and wisdom, they respected strength and skill at killing others. The weak and the poor were considered inferior to the strong and the rich.

'There were some people who did not think like this. They said: "Look at what a wonder the world is! Marvel at it and do not spread death and destruction, do not violate its natural order. We should be grateful to this earth and to her creator. Let us have reverence for all living creatures!" But most did not listen to these people, for their teaching went against the desires of both those who were thirsty for power and those whose hearts were full of fear. These fearful people constantly praised each other as great and this began to shape their deepest thoughts. They lost all humility and raised many monuments to the greatest among them. They gave titles of honor to those who had seized the common lands for themselves. They covered their warriors with precious ornaments and made metal helmets to protect and honor them. The foundation of their lives became defense and attack. This caused great harm to themselves and others and continues to do so today.

'Finally, they began building the first metal creatures for killing. Today, the whole world is infested with them. So many people have been trained to serve these people and to think in this way that humans are now lost and cannot find a way out of the great misery they have created. They never stop speaking of the need for peace, but their actions continue to serve death.

'No other creatures of the earth possess such vanity and pride. When animals kill, they do so to eat or protect themselves. It may seem cruel when we see the lion devour the antelope, but this death is part of the great order of life, and life is exalted by it. This order escapes our understanding, but it has its own harmony and necessity. When human beings appeared according to the will of the Great Designer, surely he hoped that they would honor and admire the wonders of this beautiful creation, so full of colors, odors, sounds, forms, and tastes. But instead of seeking their proper place in the midst of this creation, humans decided to try to control the miracle and make it serve them. They have spread poisons everywhere, corrupting the breath, the water, the earth, the light, and the skies. The eagle, flying alone high in the skies, sees an earth aflame with destruction. He sees humans looting their own treasures, destroying their own heritage. He sees some living in such great abundance that they have lost the ability to find satisfaction within themselves. He sees great crowds of others whose misery is so great that they cannot even feed their children. And he sees the dry, desert lands spreading like leprosy. The eagle, looking on all this from the skies, knows that this order of things is wrong and out of balance.

'All over the world humans have set up their poles with colored cloths and claimed an invisible barrier around each of them. Each group says: "This great parcel is ours; we must defend it against those outside of it. We will die for our colored cloth; we will praise it with hymns, for these are the colors that tell us who we are. We have a great leader, many fire sticks, and many men trained to use them. We also have many powerful metal creatures that will kill our enemies."

'Those who make metal creatures are sending them out all over the earth in exchange for huge numbers of cowries. They do not see themselves as criminals. They say: "We do not kill anyone ourselves with these metal creatures, for we are moral men." So they speak of morality and virtue and ask many countries to help the victims who are starving and dying because of the destruction brought by the same killer creatures they sell to these suffering lands. When these killer creatures have been used to kill large numbers of children, those who sell them claim to be horrified by this terrible act.

'The earth, one and indivisible, is being defiled by the fear, violence, arrogance, and greed of humans. Some people see this and their intelligence has awakened within them respect and humility. When they speak to their children about the world, they tell them: "This creation does not belong to us, for we are all its children. Never be arrogant toward others, for the earth, the trees, and all creatures are also its children. Live lightly on the earth, without polluting any part of it. When you take life in order to live, have gratitude. If you kill an animal, remember that life is giving itself to another life and do not waste or profane this gift. Seek the right balance in all things. Do not make useless, offensive noise, do not kill without need or hunt animals for amusement. Feel how the trees and the wind enjoy the melody they make together, and how the birds, riders of the earth's breath, are messengers between sky and earth. Be wide awake when the sun lights your paths, and when the night surrounds you, trust it. If you have neither hatred nor fear in your heart, night will lead you safely in the canoes of silence to the shores of dawn. Never worry about growing old, for old age prepares you for other births. If your brief time on this earth has been full of justice and kindness, then new dreams will be born from you like seeds of the centuries to come."

'These are the people who live in intimacy with the truth, but their innocence is powerless before the destructive actions of those who see creation as their own property. Those who treat the earth as if it were human property are always building monuments to liberty and speaking loudly about the light of truth and justice. Thus they conceal both their betrayal of these things and their skillful service to the powers that are destroying the world. They think that liberty can be built on dominating other humans and on the torture of the countless living creatures who are killed daily on their command or by their exchanges of cowries. Domination of others cannot give birth to freedom any more than a cow can give birth to an eagle. . . .'"

Tyemoro's long story had taken us very far into the night. Since I had arrived in the village, I had abandoned the habit of ever looking at a watch or clock, so I felt totally adrift. I concluded that our "civilized" habit of frequently consulting instruments that measure time amplifies the undercurrent of anxiety that pervades our lives. These instruments show only too well that our time has become a string of unrelated fragments, moments filled with acts that are often useless or even detrimental to us, and that our vaunted efficiency is mostly a compulsive frenzy that we dignify with the name "life."

In the silence the voice of my friend still resounded in my mind and in my heart. Vacillating between rage and compassion, my mind drifted into a painful reverie, gradually enlightened by the implications of his words. I realized that Tyemoro's recitation of Ousseini's words was filled with echoes of that well-educated agronomist's efforts to speak to his people in terms that could be understood by those who still live in the world of oral tradition, and who are disconcerted by their own ignorance of how the modern world works. Yet they are fortunate that they are also ignorant of the ephemeral half-truths and ideologies that have caused us to kill each other in such great numbers.

The old man was still silent, and I realized that on this unprecedented occasion he was waiting for my reaction. I began to speak in a voice that seemed foreign to me:

"Now we see that this world contains the best and the worst. The wisest people all agree that humans have gone much too far into the ways of error. We must turn from these ways back toward goodness, so that it can guide us on our way. No single group of human beings knows the way that is right for all. Most of those who claim to know such a truth walk the ways of error. Humans have learned enough of the secrets of life so that they have a choice: Serve life or destroy it. If they continue to destroy it, they are destroying their own lives and those of their children, for they themselves are children of creation. If they choose to serve it, they must not make this choice through fear, but through intelligence."

After my words, the silence washed over us again like the tide. It was as if my friend and I were drowning in it. Anyone who knows Tyemoro, however, could be sure that he was not falling asleep. I could not see that he was awake, but I sensed he was reflecting deeply, taking his time, as he always did. Finally his voice sounded again:

"Now I understand better why Ousseini chose to return here and honor the earth." Then he added, simply: "It is time for us to rest. I will tell you more when my head and my body are in harmony again."

Several days passed before Tyemoro decided it was time to continue his narrative. I spent much of this time reflecting upon the personality of Ousseini and speculating about his possible motivations. At first I suspected he might be concealing a serious failure he had experienced while abroad. I knew that many intellectuals from so-called developing countries lack sufficient financial support to complete their studies and, caught up in the sheer struggle to survive, take menial jobs. Many of them are completely at the mercy of both economic downturns in the world of the whites, and falling into disfavor with a dictatorship at home. Instead of becoming soldiers of development, they are shipwrecked in a sterile world where the simplest necessities are mountains to be scaled. It is a world with no future for them.

The glittering, deceitful logic of the technoscientific and economic model of development has shown itself incapable of living up to its promises. Yet it has not failed to spread its tentacles everywhere, draining the resources of the planet to maintain its own excess, bleeding dry vast populations through not only its greed but its lethal powers of seduction.

The reigning system of education, inspired by this logic, continues to accomplish the systematic blinding of its "human material." It now has less and less need of "material" such as students from countries like Ousseini's. In fact, by its very nature, this system cannot behave otherwise. The same is true of its own state of health. It lives on a diet consisting of a variety of presidents, politicians, all sorts of technocrats and bureaucrats, bankers, needy and hard-working citizens—and of course, the anonymous masses of inhabitants of countries without hope, living in vast cities that have lost their souls. This entire world system is devouring itself, and the digestive juices of its voracious and monstrous stomach will, in the inevitability of time, also dissolve the vanity and arrogance that created it. Often only the inevitable majesty of oblivion seems to offer some sense of nobility to this ephemeral tumult.

After having duly entertained the hypothesis that Ousseini's return to his village was motivated mainly by his need to find employment, my assessment of this agronomist began to shift as I carefully considered the details of Tyemoro's account. Now I began to suspect that Ousseini was one of those curious mixtures: a person who embodied a true marriage of modernity and tradition. These rare hybrids have the capacity to find a new path through the maze—one that avoids the blind alleys of both false progress and anachronistic values. Having freed themselves from the hypnotic trance of the dominant model, taught them as absolute truth, they clearly see its futility and its certain demise. They are those who are able to strengthen their dreams with practical knowledge without renouncing the mythic power of the dreams of their people. Only this kind of imagination can give birth to the new dreams that nourish a viable future.

This dissident intellectual began to fascinate me, and I vowed that I would meet him someday, but first I needed to hear more of Tyemoro's version of him. I knew that if I went too soon to visit Ousseini, he could not help but speak to me in my own language, offering conventional arguments meant for a European. He would surely use the familiar terms and codes that are obligatory when one educated man speaks to another: they are admirably objective but stripped of that quivering truth that reveals the human heart behind the words.

So the time arrived when, at the end of the day, I found myself once more with my friend, who sat crouched in his customary posture. At first Tyemoro spoke very slowly, as if to measure each word before finally entering into his rhythm.

11
Return to Mother Earth

"Ousseini said: 'I have understood something important. Men or women, who have no earth of their own, have no roots. They wander here and there, and when life requires them to stay in one place, that place finally begins to feel like a trap. They feel chained to it, like an animal chained to a post. Their place has no real meaning for them, and they are always ready to go somewhere else. They may travel all over the world and believe they know the world. But they still do not know its foundations. They walk not upon the living earth, but in a nameless space. This is why so many are now moving away, far from the place of their birth. When they make a life elsewhere, working, getting married, and having children, it is as if they are in a hurry, for they have no true nest. And when their children grow up, they depart as their parents did before them, leaving their fathers and mothers lonely and forsaken, waiting to breathe their last breath.

'These times of great movement of people are not natural for humans. The big villages grow larger, swallowing more and more people, and our villages are left empty in the middle of the bush. The big villages cannot continue to swallow humans forever. Even now, they cannot feed them properly and many of the people there have no work, no houses, and no medicine to keep them healthy. This is fermenting like a brew of poison.

'Many of these people who have left their homes are dying of hunger in the big villages. Some of them feel great sadness for the places where they were born—they believe their homes must have lost the favor of heaven, for the skies no longer give water there. The sun has burned away the last of the earth's saliva and even the wells and the springs have begun to fail. The wind blows the dust to distant lands and the dying earth is covered with sand. Sick animals roam the land in search of the rare living plants they can eat. This situation feels like a great, silent anger. The circle that once united the earth and its plants, animals, and people has been broken. All design has been eliminated; even tribes that were once allies turn against each other in violence so as not to die.

'Many of the wise say that humans are mostly to blame for this destruction, for they have not respected the Great Design, and have not respected each other. Others think that the whites have achieved order in their countries and have succeeded in their lives and that we would do well to imitate their ways. This is partly true and partly false. In their countries, whites suffer from the misery and loneliness of the heart just as we suffer from the misery of hunger. They speak of equality and brotherhood, but the number of suffering people will soon be great in their countries too.

'Each people has its own strengths and weaknesses. Each people must find its own path. Then they can exchange their best with each other for the good of all and the whole earth can be healed so that the eagle, flying high in the skies, can once more look down with pleasure at human beings, who are happy to be alive.

'I have spoken to you of a world about which you know nothing. I have described its workings to you so that you will understand my resolve. My initiation prepared me for a life to which I could not be true and that could not be true to me. It was a way that teaches us not to find our place in creation but to take things without limit, to profane things without limit, and to ruin the good things of the earth without limit. The wisest know that the only way out of this is to repair the broken circle by all of us working to heal the lands in which we were born and to repair the damage caused by our own wrong actions. In order to do this, we must learn the great virtue of seriousness: only if we are serious can we know peace in our hearts and know true equality among us. Without equality, there is no future for humans.

'Now, let me return to the actions that have made me firm in my resolve. For a very long time I suffered from the rejection of my own people. Some refused me any help, some turned away at my approach, and others continued to condemn me for scorning the effort and sacrifice that my initiation had cost my people. Yet I never felt any resentment toward them. I understood the grief felt by people who have had to live with the threat of the growing desert and the destruction of their community. They hoped that if I became an important man in the big village, I could help them. They believed I could never help them by returning to share their fate! Others could not accept my transgression in living in the forbidden lands and feared that the whole village might suffer from the punishment that I deserved. Only my mother, in her secret heart, kept faith in me. In silence she blessed me and prayed for me.

'My chosen land was no longer fearful to me. On the contrary, it became my greatest refuge and my only companionship in my solitude. I feel more and more certain that for vast numbers of human beings, there can be no salvation except by returning to the earth.

'Meanwhile, amidst all this contemplation and the feelings that inspired me, my chosen land was offering me plenty of hard work. I cleared a parcel of earth, broke it into good soil with my tools, and planted many seeds. The first harvest of grain was a gift of welcome. It was a good omen, for the harvest was very abundant. The earth itself was encouraging me to stay and take root like the trees that already lived there. I worked the lower lands near the swamp and used the clay I found there to build a house for shelter. I caught game in traps and discovered that there were fish in the swamp, as well as wild fruits around it. These gifts helped me to survive. I realized that I was living as my distant ancestors had lived. Yet the land around them was rich and fertile and ours is poor and tired. It has been wounded by our fires, our destruction of trees, the coming of the great drought, the rare and violent storms, and too many animals grazing on the little that has remained. We no longer know which of these evils came first and which were caused by the others.

'If we are serious about using our intelligence to heal Mother Earth, we must understand her suffering, for without her health, no creatures can live. She is the first of all living creatures, and if she dies, all creatures die.

'As I told you before, this mother of all life was born from the labor of rock and water, heat and cold and breath. Her life is also the work of countless numbers of creatures too tiny to be seen by the human eye. The first visible plants, called lichens, spread over the rock like a living carpet and gave birth to the first vegetation, which penetrated the stone, breaking it up to make food that was useful for its life and growth. The grandfather sun continued to shower his blessings of heat and light, and the new plants had plenty of water and breath as well. All of this took countless ages and endless patience. A great force was working, very slowly but with amazing power. Other, bigger plants finally appeared and fed on the invisible creatures mixed with the dust of rocks. In this way, a very important principle was established: everything that dies becomes a source of new life.

'So there came to be plants of many kinds, which grew and spread and reproduced their own kind and died. Their dead bodies, mixed with the dust, became the source of food for new creatures—insects, animals, and plants, some visible, but most of them invisible to the human eye. This life became an enormous constant work of digestion. It was as if a vast stomach spread itself over the earth and many living and dead substances were mixed to produce a new, living substance. This substance changed the earth and grew so vast that it became the food of ever more creatures too small to be seen by our eyes. All of this great work was accomplished in silence. Little by little, mixed with the remains of more and more dead creatures, this substance, full of life, joined with the clay of the mother rock, water, cold, heat, breath, and the roots of plants to make what we call earth, our soil. This great wedding feast of Mother Earth was the work of countless generations of plants and animals, both visible and invisible.

'Among these animals, earthworms and their work have a very important place. They are both made from the earth and make more earth. They are both daughters and mothers of the living, nourishing soil. This reddish-colored worm has no limbs and no apparent organs. Its body is shaped like a spindle, thick in the middle and small at its two ends, and it moves slowly, like a serpent. Some worms are very long, but most of them are not much longer than our fingers. They do not like sunlight or open air, living instead inside the flesh of the earth; they are found as deep down as the height of a man. As they move, they swallow earth and expel it behind them. This expelled earth is mixed with substances from their bodies and sometimes we can see traces of the result on the surface of the soil, like little piles of crumbled clay. Also responsible for eating the remains of plants, the earthworm's action is part of the earth's own digestion. Wherever these worms are plentiful, the earth is very fertile.

'We must understand that all this work that has created the most living part of the earth—humus, either black or brown in color—is the foundation of our own life and that of all creatures. The more the amount of humus increases, the more plants live and die in it, the greater the vitality of the earth grows, until finally great forests bloom.

'When we walk in the forest, we can see that the ground is covered with decaying leaves, branches, insects, and the excrement of worms. If we want to understand a little more about the earth's design of all this, we need only lean down and push aside the newest material. There we can see what has become of the deaths of a year ago and of many previous years. There we will find a living substance that is always changing and creating new fertility and life. If we dig deeper, we can see how all of this rests on the foundation of the earth itself, forever mixing with it and giving it a rich, dark color.

'If we dig as deep as the height of a man, we find the earth becomes hard and dense. This is earth that has not yet come alive. The earth above it is looser, broken up by the roots of trees and other plants. And above this we find the black soil. In it water and breath can move easily, with no obstacles. This earth drinks and breathes like a healthy creature.

'Humans who consider these things are closer to the secrets of life. From this kind of earth grow the most healthy and vigorous plants, and great trees reaching into the skies. Animals prosper there, and the variety and number of animals and plants are so great that no human mind can memorize them all.

'The forest is like a great cauldron cooking with a long, slow warmth. Its stew is always new, with constantly changing ingredients. The dark earth resulting from this cooking provides both fuel and the fire. A handful of this black soil is soft to the touch and has a pleasant smell. It is like a perfume penetrating the whole forest; its odor fills us with a sense of well-being.

'The earth is a living creature. The mother rock is her bones, the clay and sand her flesh, and the good dark soil her blood. This is the Great Design that has been working since the beginning. In it, fertility is born from sterility. In all creation, there is nothing that cannot be broken down to be useful. When we understand this, we begin to get a better idea of how we can heal the earth.

'The trees that reach up to the sky are a bridge between heaven and earth. These great beings are attracted by the light and heat of the sun. They spread their branches into the breath and light and their roots spread into the earth below. Unlike animals and human beings, trees and most plants cannot move. To survive, they need an earth that is rich in food and they need water to help them absorb this food. They need sunlight and heat to grow their leaves and, like all creatures, they need breath. So the tree that grows from the earth toward the sun offers us a picture of all four of the elements of creation at work, those we can hold in our hands and those we cannot. We can hold a handful of earth, but who can take a handful of sunlight, breath, or heat? The tree shows us how creation is ruled by all four of these elements.

'Like all other living creatures, trees breathe in and breathe out. The water that they take from the earth through their roots moves up through all their branches and leaves and evaporates into the sky. In this way, a great forest deposits so much water into the sky that it helps in the birth of clouds from which rain falls down to the earth to be drunk by each tree's roots and sent up again into the sky. This movement is of great importance; it maintains the abundance of the forest and all the creatures living in and around it.

'When trees are destroyed by drought, fire, or men, all of this abundance decreases and begins to disappear. Our elders know that our lands were not always the dry, half-desert we have now, scattered with a few sick and dying trees. The death of trees is a cry of pain from the earth, a call to human beings. The suffering of the earth becomes the suffering of all creatures. Our suffering from poor harvests and hunger is the suffering of the earth passing through us. We are not separate from her; we were born from her belly, our mouths are her mouth and our arms are her arms. We are akin to her most sensitive nerves. We are particles of the intelligence of the Great Design and our ability to think is a gift to the earth. But when we have been hungry for too long, this ability begins to fade and the animal in us takes control, thinking only of survival.

'We can greatly increase our understanding by contemplating the element of water. It is found almost everywhere. The cold turns it into small pebbles high in the sky, which fall as hail. Heat turns it into vapor. Between extremes of heat and cold, it is liquid. The Great Designer filled all living creatures with water. Without it, nothing on earth can live. When our body dies and decays, water begins to leave it until nothing is left of us but a dried-up husk less than one quarter the size of the living body. When we feel grief, water flows from our eyes. When we work in the heat, it drips from our body as sweat. We are made mostly of water.

'Water fills the vast basins of the earth, forming the seas. These bodies of water are so deep in some places that we do not even know what mysterious creatures live there. The heat of the sun evaporates water into the skies, where it forms clouds and returns to earth as rain. Once on earth again, it rushes down mountains, flows out of springs, and gathers itself into streams, creeks, and rivers. It hides deep inside caves, where it lies still in total silence. When there is a flood or a storm from the sea, water roars like a herd of angry buffalo. Yet it can also be so calm and clear that its surface is a mirror in which we can see ourselves, the sky, the trees, and the hills. When we hear the murmur of a spring flowing among rocks, we hear a language that consoles our heart. But when we are suffering from great thirst, water haunts our minds like a demon. When we feel it flow over our naked bodies, we are filled with satisfaction and pleasure, and when our bodies are tormented by fever or are tired and covered with dust and sweat from hard work, pure, cool water is the greatest of all gifts. Our canoes glide over its surface and our ships sail the seas in search of adventure, knowledge, or trade.

'Water is the great servant of life, but it is also the servant of death. Its anger can destroy our homes, carry away our earth, and drown us. Water corrupted by the excrement of humans and animals or by decaying bodies can become stagnant and dangerous. Certain kinds of invisible creatures multiply there, and when they penetrate our bodies, we become ill and may even die. Much human illness is due to fouled water. A pond that is too still and silent can conceal disease and death in its depths. With water, as with all things, we must use our intelligence. In it, as in all things, life and death are mixed together.

'Another element we must consider is breath. According to some traditions, it is the first of all elements, coming directly from the mouth of the Great Designer. After a mother pushes a newborn child from her belly, breath fills its breast, the beginning of the rhythm of breathing that will accompany the child its entire life until it breathes its last and dies. Along with the beating of the heart, the breath is the foundation of our body's rhythm. The rhythm of the heartbeat and breath continually speak to us of the origin of all life in the rhythms of the day and night, the rising and setting sun, and the movements of the moon and stars, for all of creation is rhythm.

'Breath also fills the space around us and penetrates every empty place, even the smallest. We can feel it as a gentle, pleasant breeze, or an angry, raging wind. The birds spread out their wings and float in breath, and sailboats capture it to move faster. It can either feed a fire or kill it, and it can either slowly dry up a pond or make its waters churn with violence.

'Breath can be warm, lukewarm, or cold. It can dissipate smoke and carry odors. Like water, when it is corrupted, it can ruin our health. Yet without breath, we cannot live. It moves rain clouds and brings hope to the farmer's heart. Yet as wind it can also steal away the earth from the farmer, darkening the skies with dust and sand, ruining the fertility of the land, and creating deserts. But when we hear the song of breath in the forest, playing the trees like flutes, our spirits rejoice at this music. Like all four pillars of life, breath is a beneficial element that can become harmful in excess.

'We have talked about the elements of earth, water, and breath. Let us now speak of light, which we also know as fire. Light, or fire, comes directly from the creative principle itself. The sun is the king of fire, traveling through the sky like a great blazing furnace. The rays it sends to earth are light mixed with heat. Even more than earth, water, and breath, light governs the great variety in all living things and assigns each its place. Countries on earth may be cold or warm, depending on how directly they receive the sun's rays. The sun shines on all creatures alike and does not care whether they are powerful or weak. It belongs to all of us and to none of us. Humans are the only creatures who have learned to use fire, to strike stones or rub wood to create it and to capture it in all sorts of ways. But fire is also found in the storm that splits the skies with thunder and deep inside the earth. For humans, fire is the most beloved and the most feared of friends. It serves us by lighting our nights, keeping us warm, and cooking our food. It has long been the servant of those who work metal, the powerful blacksmiths who are either heroes or villains in the stories of many peoples. These men make the metals we use to work the earth, the tools we use in daily life, and jewelry and ornaments. They also use fire and metal to make the weapons of violence and war.

'Without the mastery of fire, life would be far more difficult for us. This mastery enables us to find the balance between heat and cold that is warmth, thus it is a ruling principle of our bodies and those of other creatures on earth. Our bodies are like houses in which the fire has been adjusted and fine-tuned and this balance is vital for us. If the balance shifts to cold, we suffer, and if it continues to shift, we die while our bodies become cold and stiff. In some illnesses, it shifts toward heat, creating fever. A high a fever can bring death, though a moderate fever can purify our body and help it to heal.

'Humans have always been haunted by the spirit of fire. It can be the most devoted of servants or the cruelest of masters. All creatures both fear and are fascinated by the power of this spirit. When fire grows into a great blaze, it destroys our forests and our houses. Yet it can also blaze in joy, lighting our festivals and dances.

'Inside our homes, an open fire cheers our spirits and fills us with tranquil dreams. When the fire of dawn breaks in the eastern sky, it proclaims the triumph of light over darkness. Many tribes have worshipped the sun in gratitude and celebration, for they recognize its power to give us life, and dissipate certain dark fears.'"

Tyemoro paused and hesitated before continuing. He said modestly that he might not remember correctly all of Ousseini's words. I knew this was unlikely, however. Finally, I understood his hesitation when he asked me:

"Do you really want me to tell you all this? Much of Ousseini's teaching comes from what he learned from the whites, and you are white. You must already know these things. Perhaps you can now give me an opinion useful to our community. I do not need to tell you all the details of what we heard from him and did with him."

But his story had captured my interest and I was keen to hear the rest of this translation of foreign knowledge into terms a traditional Batifon could understand. I was also discovering that Tyemoro's recital contained an important teaching, in spite of my familiarity with its scientific basis. My training, which had made a specialized linguist and ethnologist, had not given me a truly universal education. As a city-dweller surrounded by concrete and metal, I realized I lacked the knowledge of people who live close to the earth. I expressed this to Tyemoro and asked him to continue and to omit nothing. In addition, from the old man's verbatim account, the mysterious character of Ousseini himself was beginning to emerge. I was now convinced that he was far more than a simple agronomist returning home after his city education: he was both a practical teacher and a kind of prophet to his people. No doubt he had suffered from the confusion and torments that arise from straddling two radically different cultures, like a man trying to ride two horses that strained in different directions. Yet I sensed that Ousseini had a firm grip on things. His achievements provided impressive proof that he was a rare kind of human catalyst. I begged Tyemoro to tell me everything. Before visiting Ousseini, and seeing for myself what kind of man he was, I wanted Tyemoro to complete his version. I was also deeply moved by the fact that this venerable elder, my mentor in the culture and ways of his people, this man who had been a friend and father to me, had unhesitatingly submitted to this new initiation by a younger man, as if assuming again the role of a humble adolescent.

He resumed his account of Ousseini's words.

12
Old Stiri's Dream

'**N**ow we see how important it is to understand the workings of the four pillars of life so that we can rebuild our earth, by relying on their power.

'Two years passed, while I worked hard on my land. I was able to survive, but I was not making progress. The criticisms of my people had softened much over time, and occasionally the few who were no longer afraid to come to my land offered me a little help when I needed it. But sometimes I became discouraged.

'One day, overcome by sadness, I went to the big village to escape my loneliness and see some old friends. Some of them were impressed by my resolve and envied it—they complained that after their long initiations similar to mine, they had lost all their own ambitions. They had grown weary of their lives, and lassitude weakened their spirits. Others mocked me, much like the people of the village. After several days there, a small number of friends became interested in my work and decided to return to my land to help me. Their arrival caused quite a stir among my people, but not for long. After a few days they calmed down, for my friends showed that they were people of integrity and the villagers soon accepted them with esteem.

'There were five of us: three men and two women. Although we came from different tribes, we knew each other well, for we had learned to read and write together. I was the only one among us who had been initiated in the study of farming. Their companionship made me very happy, for their respect for me convinced my own people that I had not made a foolish choice.

'They told my people: "We have all spent many years in the houses of initiation, but we have not learned how to help our country rebuild. Now we are seeing misery and famine everywhere, in the big villages as well as in the countryside. In years gone by, our land was able to feed all its people. Now other countries must feed us and the whites send us food—sometimes as a gift, and sometimes in exchange for whatever they may want from our lands. The big chiefs of our country are seizing most of the land's goods for themselves, exchanging them for cowries in great quantities, which they hide in the countries of the whites or exchange for metal creatures and other items made by the whites. We now understand that a people that cannot feed itself loses its liberty. The colored cloth that is the sign of our country waves not in the air of liberty, but of servitude and slavery disguised as liberty. A people or a tribe that cannot feed itself loses its dignity as well as its freedom. This is why we have rejected the illusions that our initiations put into our heads. We must use the knowledge of these initiations to free our country—and we must begin with the earth, which is the source of our food. It is the only foundation for action that will help our country and ourselves. We must also give up all our quarreling and the madness of tribe against tribe and country against country. We must stop spreading fire sticks among our people, for they bring us only ruin, and the great metal creatures of death lead only to massacres in which everyone loses. Our Mother Earth needs peace, not war, if she is to live and flower again and if her children are to live and flower."

'The villagers heard the wisdom in these words. Among the visitors was a woman who had been initiated as a healer in the ways of the whites. Another man knew much about the workings and rules governing the new world. There was a man who was skilled in the use of many metal tools and had useful knowledge about building houses and shaping wood. Another understood the workings of metal creatures.

'In a short time, our group became very strong, for we all felt that we were helping the earth, the people of the village, and ourselves. Our resolution caused the villagers' reserve to crumble. The young people still living there were delighted and they helped us with great energy. Chief Moulia himself gave us what aid he could. One day, Stiri, the most venerable elder of our community who was greatly respected for his wisdom, decided to speak to all the villagers, and everyone gathered for the occasion:

'He began: "I was troubled by the actions of Ousseini, just as you were. Now, I know that his heart is pure and a home for truth, as are the hearts of his companions. We must begin to see things as they do and do nothing to hinder them. Their actions will be of great help to us; we must also help them, for a living tree must be allowed to offer its fruits. Let them open new ways for us! They have been able to marry the white man's knowledge with the knowledge of our ancestors. Today, we are a people stranded and lost on the banks of a river with no name. Their union of two kinds of knowledge will give birth to new stars, which will guide us home, and show us how to make our land live again. As for the forbidden lands, let us fear them no more. They have been purified, and the djinns who live there have agreed to share them with us. If this were not true, terrible punishments would already have fallen on Ousseini and his friends and on us all. I know that the days left to me are few. The ancestors are already waiting for me. Before I leave you, there is something I wish to do, which was commanded in a dream. I was told that it must be done today."

'Then old Stiri stood up with difficulty. Holding his staff, he waited for a moment to calm his breath. Then he began walking down the path leading to the forbidden lands. His journey lasted a very long time, for he walked slowly and refused any assistance. When he finally crossed into the forbidden lands, he continued until he had reached the center of them. Most in the large crowd accompanying him were astonished. A few of them halted, for they were still afraid to enter these lands. But little by little, everyone gathered around him, for he was the most revered of our elders.

'He asked a child to fetch a hoe from my house. Then, slowly and painfully, he took the hoe, bent over, and broke the earth with it, as he had learned to do in his youth. When he had laboriously opened a row a few feet long, the old man took a small leather sack from a cord hanging around his neck. He poured some grains of cereal from it and scattered them carefully inside the row. When he had finished planting them, he said: "Now I have joined these young people in farming this land. Death will probably come for me before I see the harvest from these seeds. But when you see it, you will remember that old Stiri, weak and near the end of his life, was not afraid."

'From this day on, it was clear that a major change of mind and heart had swept the village. Once again, I was a full member of the community. My people offered me the greatest devotion and I offered them the same devotion in return. At last, we could unite our forces to struggle against the curse that afflicted our lands.

'A short time after this, I married Bilila, the woman who was initiated as a healer in the ways of the whites. At first, some of my people disapproved, for she is from a different tribe. But this did not last long. Most of the women were overjoyed to have a healer who could cure them and their children of many illnesses. Bilila also taught the women how to prevent illness through proper attention to clean food, pure water, healthful personal habits, and many other methods she learned from the whites and married with those of our own traditions.

'All fear and suspicion had left the villagers and a great calm now filled their hearts. They came to me to learn the ways of healing our sick earth and were willing to follow all my instructions. Abinissi, the master of lands who had refused me before, was now completely on our side. We assured him that our new ways had nothing in them that would go against his authority or the spirit of the lands for which he was responsible. We strongly encouraged him to restore the old custom of protecting the sacred forest according to the ways of our ancestors. This came as a happy surprise to him, for he thought that our white initiation had taught us to abandon our ancestors' respect for forest spirits. On the contrary, we were glad for any incentive to protect and enlarge the remaining wild areas, which are the source of game, medicinal plants, and the trees that give life to our Mother Earth and to all creatures that live near them. If these places are not protected and seen as sacred, humans sooner or later abuse them, which harms our children and our grandchildren. These forests are part of the Great Design and are a gift that must be respected and shared equally by all, today and tomorrow.

'These times freed our spirits as never before so that we were able to take on the hard task that faced us. In order to act rightly, humans must first understand the reasons and motives for their actions. This is why the initiation of my people is based on the principle of the four pillars of life. We have thought very long about our land that is threatened by desert. To heal it, we must restore the sacred circle linking earth, plants, animals, and human beings. Each needs water, breath, light, and warmth, which shows their kinship. In our land, we know that warmth and light are abundant. But water is much less abundant than in the time of our fathers, grandfathers, and ancestors. The ponds are drying or are gone and the sky no longer brings enough clouds and rain. Some of our wells are completely dry. We have noticed that when rain does fall from the sky, it washes away the good earth into swamps and rivers, where it is lost to us.

'We have noticed that the wind also carries away the earth as dust. This constant loss of our mother's blood, the earth that feeds us, must be stopped as soon as possible. In some places, only Mother Earth's bones are left in the burning sun, and nothing prospers on those rocks. The earth is baked hard by the sun so that its doors are closed; it can no longer receive the water of the skies. It should not surprise us that our harvests are worse and worse. An earth that receives water from the skies is like a full belly; we see this when we dig wells. Digging a well in good earth, we often find abundant water very soon. Sometimes, though wells may be very deep, we find very little water. Sometimes we find no water at all, no matter how deep we dig and how hard we search. The water in the bowels of the earth is not equal in all places. In some countries farmers never have to worry about their supply of water, while in other countries there is so much water that it is not good for the earth—this abundance requires a different kind of farming.

'We live in a land of very little rain and must do everything to keep it from running away uselessly. Using our knowledge of the slope of our land, we have built dams to stop the water and hold it so that it can seep into the thirsty earth. Then, using hoes and other tools, we have used the earth itself to build long ridges about two feet high, stretching horizontally along the slopes in rows from bottom to top. Sometimes we use stones placed close together to change the direction of the water when it rains. The whole village, men and women, came to help us in this work. Only the elders stayed home to take care of the children and the houses.'

"At this point in his teaching, Ousseini stopped and invited us to come with him to visit these lands. By now we were eager to see these works instead of imagining them, so that we might better understand his teaching. We followed our initiator to these lands and climbed the slopes with him. We could see immediately that a very great work had been done here: Everywhere around us, the hills were ringed with rows of earth or stone, like necklaces on the land. These followed carefully the ups and downs of the terrain, thereby causing rainwater to accumulate all over the slopes so that the hard earth would have time to absorb it.

"As we walked, Ousseini began to speak again: 'After this work, the harvests from these parcels were more abundant than those from any of the others. There is no better place for storing water than inside the earth. It penetrates deep down after a rain, and when the soil becomes dry again, the water moves up toward the surface. When we farm lands in this way, our plants can drink twice: first, from the rainwater we trap, and then from the water the earth has held deep down in reserve, which moves upward to the roots of the plants in difficult times. Any abundance of water flows down into the bowels of the earth and fills our wells. We have seen some dry wells come alive again, and even a few of our dried-up springs have begun to flow again, offering us their pure water for the first time in years.

'When the ponds and swamps fill too quickly after a rain, it means that we have not been able to trap enough water higher up. In our land, we never rejoice at seeing full ponds, for most of that water will be stolen by the sun. And when the ponds disappear in the dry season, we find nothing but hard earth, much of it washed down from the slopes surrounding the ponds. This means lost earth as well as lost water. To help even more in trapping the water, we have planted on the ridges certain herbs with very strong roots. These plants spread and protect the ridges from the effects of the wind and the footprints of animals and humans. They allow us to trap more water and prevent having to rebuild the ridges very often.

'It was good to finish this work, but we knew that there was still a great deal of water escaping down other slopes and through small valleys. So we decided to gather a great pile of stones to build a large reservoir. After digging the earth in many parts of these slopes, shaping and preparing them, we built walls all over them using the stones we had gathered. These walls had to be high and strong enough to hold back and contain the water. After the first rain, we saw how useful this was. Now we had a very large quantity of water to use for the gardens in the dry season. The reservoir also replenished the wells around it.

'All kinds of plants grew in the area around the reservoir. We even brought in some fish, and they thrived, multiplying and offering us another source of food.

'We see our work as a means of reconciliation between earth and water. They had become strangers—even enemies—to each other, but our work helped them to become friends and allies once more, increasing the fertility of the land and the good of our community.

'We all know how the constant winds of the dry season continue to ruin our lands, carrying the earth away as dust. Thinking about these lands, we saw that they had lost all protection against the element of breath, which had now become destructive. In our ancestors' time, the earth was protected by a great covering—a coat—of trees and plants, and the winds could not hurt it. This coat also protected the earth from the violence of the rains. Now it has lost its coat and lies stretched out naked everywhere around us. We decided to do something to change this.

'Our elders have taught us the directions from which the winds most often blow. We had to find a way to create barriers to lessen the force of the wind, for they also dry up plants and fruit trees. Because I had seen it done elsewhere, I knew it was possible to block the wind with living walls of brush and small trees. There are some species that are so resistant to the wind that it seems they are designed for this purpose. We discovered that our dry country has a great many plants and trees of this type: Some are full of thorns and spread their branches to the sides. Others have tightly packed branches and leaves. Nature offers many other forms and species as well.

'Any good farmer knows that the earth is not the same in all places. They know crops like cereals grow better in some places than in others, and that the same is true for wild plants and trees. Although to our eyes the earth in all places may seem the same, it is not. Different parcels favor different plants because different amounts of sun and shade and varying lengths of time that each piece of land is exposed to the sun add a different character to each parcel. We studied this, collecting seeds, cuttings, and shoots of various wild plants, planting them around the outside limits of our land, and doing whatever was necessary to help them grow quickly. This is why so many of our fields are surrounded by living walls. And because animals can also cause damage to our crops by eating and trampling them, these walls are often made of thorny or poisonous plants to keep these creatures out of our fields. In this way, the earth and plants now protected from both wind and animals can prosper and reward our hard work with good harvests.

'These are some of the ways in which we have been trying to restore the lost harmony among earth, water, and breath. We also follow once again the custom of protecting sacred woods, which now grow in more areas than they once did. All of us—men, women, children, elders, those who have been initiated in the big village and those who have not—are careful to respect the prohibition surrounding these sacred woods. From these woods come many advantages for us and our grandchildren: fresh and fragrant breezes, shelter for all kinds of life, and aid for the prosperity for our village, our crops, and our children. Our ancestors knew that saving a part of the forest in a completely wild state offers us a small part of the order and blessings of the ancient times before humans appeared.

'Now I would like to speak to you about the material of the earth, the ground on which we walk. In addition to using the earth as the source of our food, we use it for building houses and for making pots, vases, and other useful objects. Sometimes its clay can help to heal certain illnesses. And when we leave this life, our children return our bodies to the earth, where each of us again joins it and its slow process of decay. There is much to say about the earth, but let us speak of the farmer's earth, in which we plant seeds and grow vegetables, fruit trees, and all the crops that provide our everyday food.'

"Now Ousseini led us to a place where a small but deep hole had been dug in the earth," said Tyemoro. "It was about a foot wide, two feet long, and as deep as the height of a short man. Ousseini explained to us that the roots of almost all the plants we cultivate (except those of trees) never go deeper than this. The earth that had been removed from this hole was piled on the ground in three heaps. Our initiator told us to look closely at these piles, then he continued:

'This is the flesh and blood of our farmland. In order to understand it better, we have laid out these piles in the same order in which they were removed. The first mound, yellowish in color, is from the deepest part. In it we can see small, yellowish rocks. When I take a handful of this earth and squeeze it,' and he squeezed a fistful, 'it keeps the form my hand has given it. The potter can shape it into many forms and bake it in the fire so that it becomes as hard as the rock from which it was born before time and nature slowly made it into clay. Wet clay sticks to our hands and to our tools, and because it is slippery, it is hard to walk on a clay surface. Clay is able to absorb a great deal of water and hold it for a long time. It is like a reservoir for the water of the fertile earth above and keeps this moisture from escaping into the bowels of the earth far below. When we leave this clay in the open sun, it hardens and breaks into many pieces. We rarely find any plant material in it, except for a few pieces of tree roots, for it is difficult for plants to prosper in pure clay. Though it is rich in water, it is dense and lacks breath, so that its condition can easily be either extremely wet or very dry.

'The second pile of earth, which comes from the layer above the one of clay alone, is clay mixed with sand. Here, we see a little more plant material, and when I take this earth in my hands and squeeze it, it retains its form at first, but then breaks apart. Both the sand and the plant roots act to lighten the sticky density of the clay. But sand alone does not have the power to retain water, which runs through it easily, leaving it dry. Sand alone does not stick to our hands. Soil with a large amount of sand mixed in must be watered more frequently. The marriage of sand and clay is much better for plants than either sand or clay alone, but we must find the best balance. The right amount of sand makes the clay looser, and the right amount of clay makes the sand more dense.

'The soil in the third mound is from the layer closest to the surface, the one we work with our hoes and on which we walk. It is darker in color and contains much plant material. When we squeeze it in our hands, it barely retains any shape. This soil is a harmonious marriage of clay, sand, and plant material and is the most fertile type of earth. Yet it can vary a great deal from place to place, depending on the amounts of clay, sand, and plant material it contains. It will also vary according to the mother rock on which it is based, which gives the earth its character and its color. As we said before, the part of Mother Earth that feeds us is much like a great stomach. We could say that hers was the first of all stomachs and that it is really our stomach too—which means we walk on our stomach, we work our stomach with our hoes, and we plant seeds in it. This mixture of clay, sand, rock, and plant material is always digesting and making food that we cannot see with our eyes. This food is indigestible to us, but it feeds the plants, which we in turn digest. It also feeds the animals that give us milk, eggs, and meat. Thus we live inside a design founded on the earth. If we violate this design, we take the road of death, instead of life.

'Now, we must understand that even the best mixture of clay, sand, and rock will become sterile if it is not nourished. Much like a stomach, it needs nourishment in order to work and live.

'We have spoken about the material in the earth that is composed of decaying plants and animals and is digested by creatures too small to be seen. We have also spoken of the special work of earthworms, moving slowly in the earth, patiently mixing its layers and digesting plant material so that it combines well with clay and sand. All of this material constantly being digested is both the mother and the daughter of the forest and all fertile land. When its digestion is interrupted, there is less and less black, healthy soil; the earth becomes hungry, plants begin to disappear, the desert invades, and humans leave.

'Those who herd animals often do not realize this. They think only about finding plants and water for their animals but never learn about the earth that feeds these plants. When too many animals graze the land, eating every little plant they can find, they harm it. The earth loses both its food and its protection, and when it is trampled hard by the animals, water and wind are able to damage it even more. When a shepherd sees that this land can no longer feed his sheep, he cuts large branches from trees so that they can eat the leaves. In this way, these trees begin to die. The earth then becomes both widow and orphan, for she is the wife, the daughter, and the mother of the trees.'

13

Feeding the Earth

'A farmer who is ignorant of the ways of the earth is always asking her to give more food but does not feed her in return. Finally she becomes like a cow that has many calves to nurse yet has nothing to eat herself. We can all see that there is no order or justice in this.

'In the sacred circle, the earth feeds plants, the plants feed animals, the animals and plants feed human beings . . . yet who feeds the earth we farm? Only human beings can feed farmland because every year we clear it and grow new crops. Nature has no chance to feed it with the bodies of decaying plants and animals. Farmers who know the ways of the earth know that they need to feed her, just as they need to feed their livestock.'

"At this point," Tyemoro continued, "one elder said to Ousseini: 'You speak words of truth, and we agree with them. Our ancestors had ways of feeding the earth. When I was a child, we would sometimes dig many small holes, long before planting. When the winds blew, these holes became traps, holding the many pieces of plants and leaves that gathered there. They would decay, mixing with the earth, and we then planted seeds in these holes. Our ancestors found that this gave us better harvests.'

"Another person said: 'I take care every year to spread the droppings of animals, and even of human beings, on my land every year.'

"Next, a person from our own village of Membele spoke: 'For many years, I used the powder of the whites to feed the land. At first my harvests were very good, but after a while, the land became poorer than when I started. I thought I had not given it enough, so I began spreading more and more powder on it. But as time went on, my harvests declined again. Sometimes, when there was not enough rain, the plants looked as though they had been burned by this powder, and even when there was enough rain, the harvests grew worse. Finally, I had no cowries to buy powder, and I stopped using it. A number of us have stopped using the powder. We spoke among ourselves and realized that even if we had cowries to buy this powder, it was making the earth harder and dryer. That is why we decided to come here to see and learn your ways. We know that our land as it is now cannot even feed our children. The truth of your words is helping our minds to become much clearer. We see how this village and its lands have become a blessed place, a green and living country again, in spite of the desert. We bless your village and pray for it to receive the blessings of the sky. We would like to hear more about how you feed your earth. We have heard that you have new ways of doing this and we would like to learn them for the good of our own village.'

"When he had heard this, Ousseini began to speak again: 'In the traditions of our ancestors, there was much knowledge that was beneficial to the land. Yet we have abandoned those ways, and this is a great loss to us. We must restore them—but we also must realize that knowledge grows when humans share it and marry the truth of the new with the truth of the old. Today, most people of our color think that the whites have all the best ways. They turn away from the ways of their ancestors and even ridicule them, saying that proof that the whites' ways are right lies in how powerful and prosperous the white countries are.

'Certainly the whites have many teachings that are good for people of all colors, but some of their teachings are wrong. Much of their prosperity comes from taking material from other countries and giving little or nothing in return, and then selling their own goods for many cowries everywhere in the world. It is not difficult to grow fat when you are eating food from your neighbor's kettle as well as your own!

'The whites initiated me in their ways of farming the earth, but these ways have broken the sacred circle. Their own ancestors did not do this;

they respected their lands and nourished and protected them. Often, they gave special names to parcels of land they loved, as if these places were living creatures. They preserved living walls of forest and vegetation to protect their lands from the wind, the heat, and the cold. The white farmers in those days knew that their earth was alive and sensitive, but the new farmers want the earth to give more and more, and they want more and more cowries for their crops and animals. They have used great quantities of powder and poison and many metal creatures, some of which violate and destroy the earth. She is crying out in pain, but their ears are deaf to her complaint.

'By insisting on these ways, they are ruining their own food, which they produce in great abundance, but which now contains very little life force. Some of it has even become a source of sickness and death. Many whites spend more cowries trying to heal their sick bodies and minds than they spend on food. All these problems are growing worse in their countries. This path is not a good one, and we must be careful not to take it. But as I said before, even though the path their countries have taken is wrong, the lands of the whites have much useful knowledge to offer. A person who judges things as either all bad or all good has no wisdom. In the lives and actions of humans, good and evil, justice and injustice are always mixed together. In all things, we must sift out the useful and avoid the harmful.

'My last white initiator taught me that as we heal the earth, we must also purify our hearts. We must never offer our hearts as a home for resentment, jealousy, or blame, which can poison our lives. In purifying our hearts of such emotions, we purify our future actions and follow the order of the great design of the earth. Every good farmer knows that the gifts of the earth are meant to be shared among all creatures. Grains, vegetables, and fruits are for human beings; straw, and certain plants are for animals; and the decaying matter from plants and the excrement of animals are for the earth. Like the forest, farmland lives on decaying matter. Yet farmland is different, for rather than being managed by nature alone, human beings manage it and cause it to produce more. A farmer who knows the ways of nature knows that a field sometimes grows tired and needs to rest, perhaps for a year. Without this, the field may become so exhausted that it becomes infertile. Because the number of human beings is increasing everywhere, more forests are being burned to make new farmland to feed them, which is a custom we must stop. We must stop the practice of burning a field, exhausting it, and then moving on to burn a new one. This destruction of the forests hurts everyone. It is the main reason that the desert has invaded our lands. Instead, we must maintain the fertility of the fields we have by taking good care of them. We humans have the power to leave good land to future generations. Even more, we have the power to leave our land healthier than the way we found it—and we can do this while respecting and maintaining the wild forests managed by nature, which are beneficial to all creatures.'"

"Ousseini now led us to the edge of a field where there were three medium-sized piles of decaying matter. He stopped in front of one of them, pushed aside the straw on the surface, and took a handful of the dark material lying underneath. He invited us to do the same. When I held this material in my hand, I saw that it was loose, soft, and brown in color. When I smelled it, I found that the odor was quite pleasant. Some of the oldest among us laughed with delight, saying that this odor reminded them of the forests of their childhood. I too found that this smell brought back memories of my childhood, when the order of our lives was more in sync with nature. Ousseini continued:

'In your hands you hold the true food of the earth, and all of us can make this food with our own hands—but we must do it very carefully. This is the farmer's most important task as guardian of the sacred circle and the life it gives. Mother Earth gives birth to our crops, and the Great Designer intends their remains to be digested by her. Our whole village knows this now, so we waste nothing. Every family saves the precious refuse from its kitchen. The green leaves and vegetable tops we do not eat, the remains of fruits and vegetables, straw and certain plants: all go to make food for the earth. Before, people carelessly threw away these materials, mixing them with all kinds of other refuse. Bilila, my wife, is a healer, as you know. She has explained to the women that this careless-ness is not only bad for the earth, it is also bad for our health and that of our children, for this material breeds bad germs as it decays. Now our village has no more foul odors from such corruption.

'Each family takes its refuse to piles on its own parcels of land. Some families raise rabbits, chickens, or pigs, and these animals can eat much of our refuse. This they then return to us in the form of dung, which is added to the piles. Cows and goats add their own in greater quantities. We also add the old straw and stems used for their bedding and dead leaves, feathers, hair, and ashes from our fires. Sometimes we even add horns or animal bones. Like any good mother who cooks well and knows how to feed her family, our farmers and their families know how to feed their land in this way. In this mound of refuse begins digestion and a kind of fermentation that is like cooking.

'Yet if we simply spread this food upon our fields, much of it would be lost, ruined by the sun, wind, and lack of water. Even the dung of animals spread directly on fields loses a great part of its power, for feeding the earth directly in this way is like feeding too much raw food to a human being. As you will see, each family has a special place where nature can properly cook or ferment this food.

'In choosing to work this field in which you now stand, we had to consider several requirements for the "cooking" process: both water and shade are necessary, and too much sun and wind can ruin the fermentation. We have a well right over here, and where we do not have enough trees for shade, we have planted great walls of bushes or have built large haystacks. Next to the well is a trough in which we have mixed much straw with other refuse. For two days we add enough water to the trough so that the straw remains soaked. The rest of the ingredients in the pile are animal dung and other plant remains. Next to this, we have made a separate pile of clay, and you see we have collected a bucket full of ashes from our fires.

'When we select land to farm, we prefer places where water can be found. The best choice is land that has both trees for shade, and water. Some families have come to an agreement about sharing a larger common place for the fermentation of their food, which is called compost. Others use just one small place. Any arrangement is fine, as long as those creating the piles follow the rules for making good compost.

'With enough shade, the right combination of plant and animal material, and added water, ash, and clay, we can cook good food for the earth. In addition, these rows of straw you see lying on the bare earth are very good for the soil. We have dug four trenches, each four feet long, two feet wide, a foot deep, and a foot from each other. We save the earth removed from these trenches and use it to cover and eventually mix with the straw. This is another way of cooking food for the earth. Using these methods, in a little over two moons we have rich black material to add to our soil. Now, so that you will know how to do this for yourselves, I invite you to come and learn how we make a straw filling for one of these trenches.'

"Ousseini removed his robe and began to spread clay all along the bottom of a trench with a flat, iron tool. When he was finished, he explained: 'We must not forget that fermentation always requires water—but the earth will drink it too quickly unless I use this clay to help the earth absorb it more slowly. I spread the clay evenly, four fingers thick, then I wet it with three or four buckets of water. Next, I spread the animal dung and wet it as well. On top of the dung, I spread the material from the trough, which has been mixed with straw and left to soak for two days. I make this layer about half a foot thick. To help mix the layers, I walk over the straw to press it down a little, but not too much. Finally, I scatter several handfuls of ash—like adding salt to make food better.'

"When he had explained all of this, showing each step, Ousseini invited us to finish filling the trench this same way. We followed his directions, adding first clay, then water, dung, decayed material, and straw. Next, we walked on top of it as he had, and finally we scattered the ash. We repeated this, until what was the trench became a pile that rose about four feet above the level of the ground. Then Ousseini told us to shower the pile with water and, finally, cover it with earth two fingers thick. This, he said, would form a skin over the earth, making it a true stomach in which it could digest its food. We had given it the right proportions of earth and water. Because the pile was just loose enough, the breath could penetrate it, and the fire would come in three or four days.

"One of our people asked him: 'We see how the pillars of earth, water, and breath are working here. But we are very surprised to hear you say that fire will come soon, for we do not understand how.'

"Ousseini answered him: 'I understand your confusion. During my initiation I was surprised when I first saw the fire. Come and look at this pile of compost, which is eighteen days old. We have just turned it, given it its first stirring. There is less fire than there was, but you can still feel it if you put your hands in the pile."

Tyemoro said: "On his suggestion, we all did so and realized that it was true. The material felt very warm inside, even though it had never been exposed to the sun. Ousseini continued:

'In the early days of cooking, the mound is so hot you cannot hold your hand inside. It is like a human's high fever, which purifies the body. The heat causes water to rise from the compost as vapor and the different ingredients in the mixture soften, crumble, combine, and begin to give off a pleasant odor—much like a pot on the fire. Now, just as a good cook adjusts the fire to be sure the food does not burn, we must keep track of this heat to be sure not to burn this food. The high heat must not last more than three or four days. If it does, we water the entire compost heap to cool it enough so that it is comfortable to keep your hand in it. Like a fever, the heat is beneficial when moderate, but must be lessened if it rises too high.'

"One of our group asked if more ingredients could be added to the compost heap after the cooking has begun. Ousseini said no: When the kettle is full and has begun to cook, it should not be disturbed by new ingredients, though its contents must occasionally be turned and mixed. Beside this, though, once the compost has begun its fermentation, it must be allowed to complete it. Ousseini explained it to us: 'Imagine a woman is cooking rice for some guests. When the cooking has been going on for some time, someone comes to tell her that there will be more guests than expected. She knows there will not be enough rice, but does she add more rice to the kettle she is already preparing? No, for this will produce a mixture of undercooked and overcooked rice. Instead, she prepares a new batch of rice in a different kettle. The same is true in cooking food for the earth.

'We must also make sure the ingredients mix well in our compost heap, as we do when we stir a pot of stew. We create a new, empty trench beside the full one. Then we remove the skin and covering from the heap and shovel the top layers of the pile into the new trench, so they are now at the bottom. We keep shoveling so that the bottom layers of the first heap become the topmost layers of the new pile, mixing as we go until the compost heap reaches its original height of four or five feet. If the pile is too dry, we add water, for remember: Lack of water can slow down or even stop the fermentation. Too much water is also bad for the process. To find the right amount, think of a cloth that has been soaked in water and wrung out. This is the kind of dampness we are looking for in a compost pile—no more, no less. When we turn over the pile the second and third time, we must also make sure it has the correct balance of water, fire, and breath. These second and third turnings allow us an opportunity to mix the ingredients ever more thoroughly. As the ingredients crumble and combine, the pile will grow smaller. Though the warmth decreases, it is still present, as in a healthy human body. When this digestion is occurring properly, it creates no foul odors. Instead of causing revulsion, it is even pleasant to us. As part of the great design, there are many beneficial germs, too small for us to see, that multiply and perform this digestion, producing rich food for the earth, which we here have seen and touched.'"

Tyemoro paused, then said: "There was one young man from Membele who had been silent all this time. Ousseini now turned to him and asked if he had any questions or comments about what he had heard and seen. After some hesitation, the young man spoke: 'One day, some whites came to our village to show us how to use our garbage to feed the earth. They told us to dig much bigger holes than these—you could have buried two buffaloes in each. Then they told us to throw our refuse and excrement into the holes every day, occasionally water the growing piles we created and pack them down, and then add a layer of the powder of the whites. This whole procedure was repeated until each hole was finally full. When planting season began, we spread the material from these holes onto our fields. It was very different from your food for the earth: It was dark black, soft, and sticky and it smelled bad and sent an unpleasant odor through the whole village. We supposed this bad smell was a sign that the mixture was good for our fields, but now we see that the whites' methods are very different from the principles you teach.'

"Ousseini answered him: 'Yes, we have heard that the whites taught this different way in many of our villages—but theirs is not a good method. In the first place, it does not consider the difference between rotting and fermentation. The ingredients are packed tightly into a hole that is much too deep, which does not allow enough breath. Lack of breath brings death, just as it does when we suffocate or drown. Imagine that a hunter kills an antelope so large that there is too much meat for his family to eat right away, so he saves the extra meat for later by cutting it up and putting it into closed jars. When he opens one of the jars after a few days, does he find good meat? No, the meat has rotted; it is dangerous to eat and must be discarded. Lack of breath causes bad germs to prosper. When meat smells bad to us, we know something is wrong with it, and the same is true for food for the earth. This cooking, or fermentation, can happen only when there is enough breath available. If there is not enough breath, it starts rotting, which is not the same as fermenting. Land fed on good food is healthy land; likewise, the plants, animals, and human beings that live on it are healthy. The special plant remedies of our ancestors and the new remedies of the whites alone are not enough to keep us healthy. The first source of our health is the earth.'"

"After listening for a long time without speaking," Tyemoro said, "I asked our initiator what should be our first step in making our own offering of food for the earth.

"Ousseini said: 'In order to give the best answer to venerable Tyemoro's question, I would like to show you this compost pile in the fourth trench, which has almost finished cooking. Now, watch as I lift up the straw covering, and open the pile. . . . You see, this is the material we want. It is now almost cool and has no bad odor; in fact, it is quite pleasant to smell. We see that there are red worms living in it, which is a very good sign, though there are not always as many as there are here. Sometimes we find other less beneficial creatures living in the pile, but this is not a big problem. Look—here is a gray worm with tiny feet. It curls up when you hold it. We remove these worms from the compost and feed them to our chickens. While they are good food for them, there is a more important reason to remove them: if they get in our fields, these worms can damage the roots of our plants.

'So you see, this food for the earth is a living material that must be managed with great care. We must offer it to the earth at the right moment, when this food is full of life. In this way, it can restore the life and youth of the earth. If we wait too long, however, this material begins to decline and finally die, just like all life. When it feels like sand, has no smell, and no longer holds water, it is dead. But its life is long enough to allow us plenty of time to use all of it. I use one of three methods of spreading this compost, according to how much of it I have, and how big the field is. As an example, come with me to this field over here.

'As you can see, this piece of land is about one hundred feet long and one hundred feet wide. I want to grow cereal grains here, but suppose I have only one compost pile. This is not much, so I dig little holes about one foot apart in the earth of the field. Each of these holes is big enough to contain one good-sized handful of compost. In this method, our greatest enemy is the heat of the sun, which kills the beneficial germs in the compost. So I must lose no time in planting seeds in each hole and covering them with the earth I first dug from the hole.

'If, however, I have three piles of compost to use in this field, I can spread it on each of the planted rows and mix it with the earth. If I have six piles, I can spread the compost evenly over the entire field, and mix it well, using a hoe. This allows me to plant seeds in the pattern I've chosen—but I must always be careful not to press the earth, for it must breathe.

'Sometimes our compost cooks too quickly. In this way, it is ready too soon; by the time we need it, much of its power is gone, just as it is from overcooked food. To prevent this, we spread the compost in the shade too dry—but it must not dry too fast and it must not be exposed to the sun. If there is no shade, we spread it out at the end of the day, let it dry overnight, and then pile it again at dawn and cover it with straw. If necessary, we repeat this for several days to slow down the cooking. Then, when it is almost dry, we put it into sacks, which we keep in a sheltered place. In this way, we have a reserve of compost, which will be useful for all our crops—vegetables, cereals, and fruit trees. When this dried compost is spread and mixed well with the earth and a little rain or water, it regains all its life.

'Since beginning to offer this food to our lands, we have seen the earth come back to life. It is easier to hoe; water and breath can easily penetrate it. It gives us great joy that this soil holds water much better than the soil we used when we farmed the old way. In fact, it can hold three or four times its own weight in water. We can test this by weighing a bit of dry soil, adding water until it the earth can hold no more, and then weighing the soil-water mixture.

'I could say much more about the benefits of this wonderful food for the earth, but I think I have explained the most useful ones to you. It is now up to you to decide whether you want to try these methods and put into practice the principles you have learned here. Do not think of this knowledge as a special secret; share it with everyone generously, just as we have shared it with you. It is for the good of the earth and all her children.'"

14
The Initiated Initiator

Tyemoro continued, "Ousseini then led us on a walk through their fields and gardens, teaching as we went along. We saw fruit trees, vegetables, cereals, and even medicinal plants, as well as plants used for spices. The sight of these things brought pleasure to our eyes and joy to our hearts.

"Our guide also recommended that we leave as little bare ground as possible in our fields: 'The earth bearing our crops must be protected from overdrying by the sun and the wind. Be sure to cover with straw, dead leaves, or dead weeds the empty spaces between the planted rows. This helps the earth retain water.

'In addition, plants grow better in well-loosened soil. To keep the earth around plants from becoming too hard, we hoe it to loosen it, which allows the soil to breathe better and hold water. But as any good farmer knows, the soil must be watered carefully; too much water will carry the good food too deep in the earth, where the roots cannot reach it.

'I have spoken of the basic methods we need to know and understand in order to help the earth come alive again and remain healthy. There is rightness in these acts, for they restore the sacred circle: the earth feeds plants, plants feed animals, plants and animals feed human beings, and human beings feed the earth. Of course, as any farmer knows, there are many tasks to see to beyond these: there are planting, weeding, trimming, transplanting, and harvesting—everything in its time, and different plants according to their needs. Most of this, we have learned from our ancestors. But old knowledge has always been enriched by new knowledge. We must be vigilant so that we do not lose the old ways, yet also we must help them to grow.'"

Tyemoro's story had taken us very late into the night, but never once had my interest flagged. The only sound I had uttered was an occasional murmur of encouragement during brief pauses in the old man's long account. I sensed that he was still tempted to diminish its importance, suspecting that most of the story's particulars were banalities to me. Perhaps he also felt that he was treading on new ground, where he was somewhat disoriented. Nevertheless, he offered me some concluding words:

"This teaching has brought a new order of things into our minds. What we saw that day with Ousseini convinced us of its truth. We learned not only with our minds, but with our eyes, our ears, our noses, our hands, and our mouths—for the taste of the fruits, vegetables, meat, and grain that our hosts fed us confirmed the truth of this teaching: everything was delicious. According to Ousseini, when the Mother Earth lives in delight, she produces food that delights us. This is why we must give her nothing that smells foul or has been made through corruption or violent procedures."

Tyemoro made a sign that with this, he had finished his narrative, yet something in his manner suggested that he felt he had not fully completed his presentation of Ousseini's teaching.

During the next three days, life went on as usual in the village. At first, I did not understand why this elder was still waiting to hear my reaction to Ousseini's instruction—the concrete results it had achieved seemed more than sufficient proof of its soundness. When we met again, Tyemoro still seemed to retain a tiny residue of caution regarding Ousseini's ways. After all, their community had experienced great suffering by accepting new knowledge from whites—knowledge that concealed ruin behind the most attractive appearance. Because I was both white and his friend, he wanted to be sure that I did not sense a similar possibility in these new ways before offering his formal approval for Membele to follow the village of Mafi and fully engage them. I told him that Ousseini's teaching was not white knowledge alone, but instead was a marriage of human knowledge from a number of traditions. I told him that the future of humanity must be founded on humans sharing different forms of knowledge and applying them in ways appropriate to their own conditions.

To me, Ousseini's teachings appeared to be an outstanding example of this new way. He had no doubt realized that for a vital community to be restored to his people, the resources of the land itself must be revived by mobilizing men and women to live there and cultivate it through healthy means. This strategy seemed all the more intelligent to me given the contemporary reality of overpopulation and misery in urban areas, and the uprooting and wandering of whole peoples. I felt a regret at my own ignorance of the sciences of the earth. I had only my own common sense to rely on to try to understand these procedures and methods. In any case, my curiosity had reached a peak, and it was now time to make my own pilgrimage to Mafi, especially considering that my scheduled return to Europe was fast approaching.

For practical reasons, I planned my visit for my way from Membele back to Europe. Yet I was reluctant to leave until I was sure that Tyemoro was not holding back some of his words in his doubt about their usefulness to me. To be honest, the issue of the agricultural situation in Membele, grave and important as it was, was a pretext to draw out my time with Tyemoro. Our relationship deeply moved me and his words strengthened this feeling. The idea of leaving him, perhaps never to see him again, made me feel like an orphan in the world.

The heat of April was growing oppressive. At midday in Membele the sun seizes the center of the village, forcing all humans and animals to leave. Sometimes in the early morning I go to a nearby hamlet where there is a so-called market, a collection of straw huts or makeshift shelters where people exhibit the fragility of their existence more than their wares. I remember a European economist who once told me that he made a habit of inspecting both markets and public dumps in countries he visited, for these two gave him a vivid snapshot of the nature of the economies in these places. He offered, somewhere between mockery and anger: "In certain rich countries, 15 percent of all human energy and resources are mobilized to produce nothing but throwaway material. Tossing things in the garbage has become a mere reflex for citizens of wealthy countries—and then they point the finger at poor countries for having too many children. If you add to this the enormous costs of waste management, you get a full picture of the monstrosity of the situation." As for the people in the market village I visited, they could hardly be accused of wasting anything. Their garbage consisted of only very small piles, quickly dried by the sun and dispersed by the wind.

On the day when I went to this nearby village after my long meeting with Tyemoro, I found that my morning walk was touched with uncharacteristic sadness. I was more attentive than usual, as if I was carefully absorbing and storing the sights around me for later memories and visions. Faces, gestures, voices, colors, odors—all took on an increased sharpness. Once again, I found myself admiring these people whose very survival is under daily threat and who bear it with a cheerful nonchalance.

Absorbed by these impressions, I suddenly felt something small and soft slipping into my hand. I did not have to look to know that it was the hand of little Ninou holding mine. As usual, his grip was warm, relaxed, and completely trusting, with no hint of demand or expectation. I also did not have to look to know that Ninou was wearing a broad smile and a proud expression as he walked with me in this way. He was barefoot in the dust and his long, light robe flapped against his thin little calves. I had long ago made a decision not to privilege him with special gifts. He seemed overjoyed enough by my affection, which was all he asked. I gave all my modest material contributions to his community. Not that this was expected of me; on the contrary, the hospitality they offered me was always unconditional and I had to insist that they accepted my contributions. My friendship with Tyemoro somehow conferred on me an aura that engendered a certain respect and deference. As for Ninou, I decided that he should grow up here. I would indeed watch over the welfare of this seedling, but it must grow here, in its own land.

One early morning, I found Tyemoro alone, squatting underneath an acacia tree, just outside the village. He seemed plunged in deep meditation and he sat as still as a stone. My reluctance to disturb him was overcome by my desire to join him. I hoped that this importunity would be excused by the fact of my imminent departure. I soon noticed that my old friend also seemed a little sad. Sensing my hesitant approach, he made a gesture for me to sit with him. Then he spoke:

"I know this visit of yours has a special meaning. You have not asked me a single question about our language or our customs, as you have before. Perhaps you have gathered all you need, trapped in your word box and written on your papers. Instead, you wanted to know the reasons for our misery and I have tried to explain them. Perhaps they are not the whole truth, for the world does not allow us to fully understand its ways and rhythms. If you had not become my son, I would have died in ignorance of many things about the world of the whites. I have also learned much from you about your life among your own people. Soon, you must leave, and I will never see you again. This I know. I also know that your presence has helped me to reflect upon the time in which we live in and to try to do better in my role as elder and advisor to my people.

"Ousseini is on the right path. And there will come many more people like him. You know our community, how it has lost its way in a time of bitter changes. I also ask you to do what you can in the future to help us, for now you are one of us. You are even more important than our children, for you are a guardian of our traditions.

"I have decided to send one of my sons and four or five others to be initiated by Ousseini. He has generously offered to do this, and in exchange, these children of ours will help Ousseini and his people work their fields. When they return, these young will initiate all of us and help life to return here as it has to Mafi. This is my greatest wish."

I told Tyemoro how good I thought this decision was. Then, in a voice trembling with emotion, I vowed to do everything within my power to help their community. He was silent for a long time, as if absorbing my words and allowing me to measure their gravity. It had been a long time since my eyes were so full of tears.

From the village we could hear the laughter of women, the shouts of children, and the eternal brays of the donkeys. I was overcome by a sudden fantasy: Why not move here and bring Madeleine and our children with me? But this was a dream with no foundation, for how would we support ourselves? Also, I knew that my wife could not bear these harsh conditions for long, to say nothing of the children. . . . I saw that I could not escape the fate of my heart being torn.

I was still preoccupied with these futile imaginings, when Tyemoro brought me back to the present. As if offering a pretext to spend more time together in these last days, he said:

"I still have a few things to tell you about Ousseini's teaching. At first I thought they were of no importance, but last night, I was troubled by a sense of the incomplete. Because you have asked me to hold nothing back, I will bring my words to completion.

"Two days before we left his village, Ousseini said that he had spoken enough to us about farming the earth. He wanted to speak other words that concerned plants, animals, and even human beings. It seemed very important to him to do this. When we had gathered to listen to our initiator, he said:

'Plants, like all creatures, are the living emergence of the forces of earth and sky. They prosper in places that they prefer, each according to its kind. These places have the right combination of conditions of heat, light, breath, water, and earth, as well as other subtle aspects that favor one or another family of plants. These families are much like tribes for different kinds of plants. There are so many types of plants that our knowledge cannot contain or understand them all. From the smallest seed, barely visible to the eye, a tree can grow so tall that we are tiny beside it. Ever since we humans appeared on the earth, we have learned to eat the fruits, grains, leaves, and roots of many plants, and, as I have said, we owe so much of our life to trees.

'Plants, like other creatures, are adapted to the lands in which they live. Where water is abundant, their leaves are very large to allow them to breathe and grow in abundance. Where water is scarce, plants' leaves are smaller and the plants are harder and are often covered with thorns, as if they understood that they must defend against animals their special ability to store water. Some people say that plants even have their own language, which we cannot understand. Plants may be either helpful or harmful toward each other. Some plants cure our ills, while others poison us. Plants and human beings have grown closer over the centuries.

We have learned to use their natural gifts, mating different kinds to create bigger fruit or roots. Human life is inseparable from that of plants.

'In the beginning, each family of plants was born, prospered, multiplied, and died on the lands given it by nature. Though these plants could not move, their seeds could spread by wind, water, and other means. Animals and humans ate according to their needs, avoiding plants that were useless or harmful to them. In those times, nature created a harmony among living beings that was different in different climates. All beings were related to each other in the great design but each had its own timeline, founded on the fundamental principles of life and death.

'Because of the frequent wanderings of humans, who learned to grow seeds into seedlings and transplant them, plants became great voyagers. In addition, many plants used for food and medicine were exchanged for different growing things from distant lands. Tomatoes, potatoes, eggplants, corn (maize), and many other vegetables we now use did not exist in our lands in those days, and many of our own native plants now grow in distant countries. Long before the whites, the peoples of Egypt and the Middle East became very skilled in the art of growing many kind of plants. Shaping earth, water, and plants, they built vast gardens with many fruits and flowers to delight the eye.

'Our vast knowledge of so many species of plants has come down to us through countless generations of different peoples. Humans have always shared this knowledge to the advantage of all. Rather than being the property of the few, it is our common heritage. Knowledge of plants has always been a commerce of life, not of death. Farmers have been able to save the seeds from each year's harvest to continue old or nurture new varieties, growing healthy plants according to the conditions of the land. For those who work the land, this process has been a source of satisfaction and success for untold generations.

'But today, this order is no longer respected, which has caused great trouble. Many people in distant countries have developed a taste for new plants. White farmers on large farms, possessed by the desire for many cowries, have used their knowledge to change the nature of many species of crops. These new plants offer increased production, but many nourishing species that have come down to us from the patience and knowledge of the ancestors of all peoples have been neglected or lost.

The only goal of such farmers is to produce great quantities of food for their big villages, using plants that are the result of hasty marriages of species accomplished by the cleverness of their learned men. Little by little, many of these new varieties have lost the vigor of the old crops from which they are descended. They have become so weak that they can survive only with the use of large quantities of water and white powder. This powder is like a kind of salt making the land very thirsty. The unnatural weakness of these plants also draws to them insects and disease—as if these two forces were trying to establish a more natural order. To fight this, the white farmers use many poisons that are bad for the land and for humans. Some of these added materials produce special fruits, grains, and vegetables that are unable to reproduce in a normal way. Farmers who want to grow them again must pay for seeds every year. It is true that these seeds often give larger crops, for they are like mules, which result from mating mares with donkeys. The mule may be unusually strong and useful, but it is sterile. When small farmers must pay cowries each year to buy these seeds for crops, they lose their freedom and their dignity as well as all their cowries.

'It has always been human nature for people to seek exchanges that are advantageous to them, but in our time, money has become more important than humans. This is why we must take care to safeguard our heritage of seeds that can grow into healthy and fertile plants, and why we must use our imagination and knowledge to produce crops from them that live in harmony with our land and water. Many peoples have become careless of this, adopting plants that come from far away with no concern for what their own land already offers in abundance. Many people pay cowries for food that is shipped from far away and, as a result, has lost many of its virtues and power to keep humans healthy. Each people must learn to restore its links with its own nourishing earth, if it is to recover its liberty. We can also be open to riches from other lands, but in a fair exchange that respects the kinship and common good of all peoples.

'There are also countless varieties of animals on earth. They may live in water, between water and land, only on land, between the sky and the land, or between the sky and the water. They are also our ancient companions on earth, and many of them were here long before we arrived.

We have learned to feed ourselves by hunting and fishing them—by capturing them in many ways. Through this we find our place in the circle of beings who live, breathe, and die, offering their lives to this sacred circle of life. Life always gives itself to life. This principle has been the same since the beginning of time, though certain peoples, in observance of their beliefs and customs, have chosen never to eat the flesh of animals.

'Some animals are nearer to us than others. Insects, for example, are quite foreign, and because of this, it is more difficult to establish a relationship with them than it is with other animals. Fish are also distant from us: they breathe under the water, where we drown, and they drown in the open air, where we breathe. The world in our oceans is alive and strange and has many mysteries that humans have not been able to unveil.

'Birds live in the skies, on land, or in water. Some eat insects or small land animals, while others live on water animals and plants. Some hunt at night because their eyes are able to pierce the shadows. Birds can walk on two legs or fly with two wings, but some animals have no legs or arms or wings; these crawl, using their skillful twisting movements.

'Female insects, serpents, fish, and birds lay eggs fertilized by their male counterparts. Most of these creatures have either scales or feathers and are part of a large group considered to be quite different from the group made up of those whose young are born live and who drink milk, as we do, until they develop teeth to eat other food. Some in this group eat meat, some eat only plants, and some eat both.

'Like all living creatures, animals have spent countless generations adapting to the conditions of their land, which may vary from extreme cold to extreme heat to anything in between. Some have thick fur to protect them from the cold, while others have developed in ways that protect them from heat and thirst. Some are very powerful and do not need to multiply in great numbers to keep from being eaten, for they defend themselves well. Others that multiply in greater numbers may become very swift and clever at avoiding danger. In this order of life, all must reproduce their kind in order to survive, and all must die. As we said before, life and death are closely bound together, and all species work together to maintain the life and fertility of the earth.

'Humans discovered long ago how to tame certain animals to

become companions. Sometimes bonds with these companion animals are so strong that these creatures—especially dogs and cats—live with humans in their homes. Cats help humans by guarding stores of grain from rats and mice, and dogs, with their speed and sense of smell, have learned to herd and protect livestock and to help with the hunt. They can be fierce or gentle, according to need.

'Through training and through their own qualities, animals have learned to serve humans in many ways. Some, such as sheep and goats, have been bred by us for a very long time for their wool, milk, their meat, and skin. Buffalo offer us their strength to help us work the land. Chickens, ducks, pigeons, and many other birds give us their eggs, their meat, and even their feathers.

'Some peoples say the horse is their best friend, for their way of life depends on its speed and strength. Many humans have used this animal to travel over great distances in the world either to gain knowledge or to make war and steal the lands of others. The horse's speed and strength has served life but has also served greed and violence.

'Humans have spread animals around the world much as they have spread plants. In our time, whites have prompted many changes in the treatment of animals. Some people, blinded by their greed for cowries, have begun to treat creatures with great violence: they have destroyed great herds of wild game on which many tribes have lived, killing buffalo only for their skins or tongues, or elephants for their tusks, and leaving their bodies to rot. These are only a few of the great cruelties humans have inflicted on animals.

'In our time humans have become arrogant, with little respect for the earth. Sometimes they claim that the Great Designer made everything for us alone. Other times, they deny the very existence of the Great Designer, saying that reason alone is enough to place human beings above all other creatures, giving them power without limit and freeing them from nature, which has become their enemy. They claim reason can remove the veil from all mysteries. Their world is one of anxiety, discontent, and violence against one another, animals, plants, and the earth. They have lost all sense of the sacred—animals no longer appear to them as fellow creatures who feel pain and joy, as we do. Often, they see animals only as sources of meat. I have already told you about the great houses in the countries of the whites, where animals are crowded together in large numbers, kept for their meat, milk, or eggs. Cows, pigs, or chickens are packed together cruelly. They cry out and try to move, living in darkness lit only by cold torches. They no longer know whether it is night or day outside. Often, they die of fatigue. In these places many young cows are fed a special diet so that their flesh stays white in spite of their age, for white flesh brings more cowries than red. These animals, created to know the joy of moving and running, are kept imprisoned by walls and chains, where they must be motionless, like plants and where they are fed unnatural foods. This cruelty was devised by humans who are always dissatisfied, who want these young cows to grow more fat and to give their meat earlier. These creatures are born in suffering and die in suffering. Humans have left kindness so far behind that they now feed themselves on this suffering without understanding the suffering they create for themselves, for earth, plants, and animals are all linked by the same breath, the same life, the same force.

'Strangely, while some animals in such countries know only suffering, others are treated like princes and princesses. Some of the dogs and cats that whites keep are fed more and better quality food than children in poor countries. This also goes against the sacred order of things. Animals we keep should be allowed as much as possible to live according to their own nature. Humans must use the gift of their intelligence to find their true place in the order of things, which calls us to respect the plants and animals in our charge, for they devote their lives to us. But this can happen only when our hearts are free of greed. When we rediscover humility, compassion, and moderation, life will once again become a great blessing for us all.'"

Tyemoro had finished. Before parting, we sat long in silence together, basking in that familiar morning current in which the imagination is free to float, like a boat loosed from its moorings.

15
Departure from Membele

L ast night's cool still lingered in his room. Waves of the pain of parting surged through me. At times like this, you promise yourself that you'll return as soon as possible, forgetting that it is life itself that will decide. Tyemoro quickly brought me to the surface, seizing my hand in his. He spoke severely.

"Do not let emotion possess you. Emotion is an inevitable fire, but we should always be very prudent about throwing on that blaze the wood of our memories and regrets."

His sudden, forceful speech and gesture gave me a mild shock, for they came from a man whose actions are so imbued with calm and repose. But the shock was useful.

He then added: "Do not let time corrupt the grains you have stored in your memory. Draw all their strength from them and always be ready for other harvests."

Then the old man put his hand on my head and left it there for a long time. Little by little, a palpable sense of warmth, well-being, and deep peace settled in me. Then he told me: "When you think of me in the future, think especially of this moment. The voice you have trapped in your box is not the voice of Tyemoro. Listen to it as much as you like, but do not marry it with my image. Never say to yourself, 'That is Tyemoro's voice.' Instead, say: 'That is a voice.' Also, when you return someday to visit your brothers and sisters here, do not let yourself be distracted by any visible traces of life I may have left behind. Instead, be with me in that which never dies. Along with the nourishment of our ancestors' teaching, which I have confided to you, I have given you a small bowl. It is your task to put a little food in that bowl each day and share it with others, for it must be shared. If you are able to help others understand what we have lived in the secret of our hearts, your duty will be accomplished. It is good for humans to go beyond appearances in order to understand more and live better together.

"I have never left the place of my birth, and I know nothing of the great world outside. There are so many human beings on this earth, they say, and they have so many different customs and feelings. It seems to me that when people see each other as strangers, they are more likely to hurt each other than to understand each other. I do know one thing: A hand is useful only when its fingers work together. The bigger and stronger fingers are not prouder than the smaller ones. All work together and each finds its place."

He seemed about to say more, then stopped, with a little shake of his head. "Forgive me, I've talked too much! Please give our greetings to Ousseini, and tell him we are eager to see him."

The bush taxi had arrived and was waiting not far away. Tyemoro held me in a long embrace and then released me and gave me a little shove of encouragement, as if freeing a bird. My friends had already loaded my baggage and arranged it in the little truck. I walked toward the vehicle, this time escorted by the entire village, including women. In the air there was a mixture of noisy exuberance and the pain of separation.

I felt Ninou's tiny hand slip into mine and grasp it more tightly than ever before. I did not have to look to know he was in tears. Since that early encounter in my room, over the course of my stay, we had spoken a great deal. I tried to reassure him that of course I would return and that I would never forsake him. Indeed, how could I, in my heart, ever forsake that look of his, so trusting and innocent, plunging you to the depths of your own consciousness? Ninou was to me like a little sapling that needs protection so that it can grow and thrive in its own land, among its own kind. Just as Tyemoro had adopted me as a son, I sensed my duty of paternity toward Ninou. I would assume this role in an appropriate way, with prudence and discretion. As his face moved between a broad grin and tears, this parting image of Ninou burned long in my memory.

Standing on the running board of the old taxi, I now became the center of the spectacle of endless goodbyes. It went on so long that my patience was tried and I wished the driver would start the motor. Finally, he did, and we began to roll. Looking back at the crowd of children running behind, I suddenly and unthinkingly made a grotesque face at them. It had the desired effect: an outburst of laughter. Perhaps this was my own way of trying to return the gift they had given me: laughter somehow rising out of the joy of life, in spite of all difficulties. I watched the last silhouette still waving, inviting me to return again, as it disappeared behind the last hill.

My physical image of Ousseini had been formed from listening to Tyemoro: I pictured him as a tall, thin man with a resolute step. As usual, the reality was somewhat different. Before me stood a man of medium height and solid build. His gaze communicated a strong will and calm intensity. He was a man at home in his body and in his place—a man who easily inspired confidence.

The village of Mafi fully deserved its reputation. Staring at this great swath of greenery and life in the midst of a desert environment, I reflected how little we measure the miracles that can arise from generous intelligence and constructive will. Not only did this green island emanate a palpable sense of well-being, it ennobled the arid country around it. Much of its landscaping followed the principles of antierosion, and many young trees now grew there. The place was like a message of hope carved in the very flesh of death. Now I understood better what was meant by the sickness of the earth and how it is possible for humans to transform themselves from destructive parasites to healers.

Ousseini introduced me to the villagers as Tyemoro's son, which immediately conferred upon me an aura of respect and delicate attention. The next morning, he took me on a tour of their various works, accompanied by his own instructive commentaries dealing with agriculture and its relation to ecology, expressed for my benefit in technical and scientific terms.

I soon realized that the success of those in Mafi could not have been achieved without an extraordinary mobilization of the village women. In Mafi women evidently freely expressed their feelings and ideas. Previously, they had been so overwhelmed by their roles as guardians of sheer survival, with its endless tedium of daily chores to keep their families alive, that this move from passivity into active participation had made them a new force in the community. "It has not been easy," one of the women leaders told me. "But now everyone recognizes the benefits that occur when we also take the initiative."

Her group, an admirably organized one, managed the community chicken and animal sheds, which entailed seeing that kitchen refuse and produce from the land was properly collected to feed the animals. Women also worked at transforming food into preserves, at weaving baskets and cotton, and at making clothes.

The community had acquired a mechanized mill, which freed them from otherwise long and tedious tasks. A small cooperative business had formed, enabling the sale of certain products. In general, community resources were available to put toward whatever was needed: tools and facilities, stock for the village pharmacy, certain repairs and improvements, and so forth. This general principle relied first and foremost on human energy, but animal energy was also often employed. A deliberate choice had been made to avoid machines and use animals— oxen, donkeys, horses—for power and strength. As villagers explained to me, they did not exclude the use of some machines, but such use was kept to a minimum and top priority was placed mostly on the solidity and dependability of the vehicle to minimize the risk of costly repair. They had learned the lesson of overdependence on industrial products from the north and knew all too well the danger this posed to their resources.

"Lately," Ousseini told me, "a number of agronomists, researchers, and journalists have become interested in us. There are even university students who want to write their theses on what we are doing. It appears that we have become a kind of case study. Yet we remain vigilantly faithful to our principles and we have plenty of responses to offer those who want to reduce us to mere tinkers, experimenters, or worse, some sort of cult."

The first meal in my honor was attended by a doctor, a nurse (Ousseini's wife), an economist, a mechanic, and Chief Moulia. Most of the people at the table could be called the "intelligentsia" of the village. All were clearly of one mind in their joy at finally having a sense that their gifts were serving a true community. None had the slightest doubt as to the choice they had made.

In a long aside to me, Ousseini said: "I don't know anything about your beliefs, but there are a lot of people who think that this deep and indefinable feeling of the sacred we have within us is a dangerous delusion, a weapon of mystagogy that is bound to lead to the formation of cults and to intolerance of others. But we have no dogmas—unless respect and reverence for an order of life confirmed by science and common sense is considered a dogma. As for our rituals, perhaps they arise out of ancient gestures beyond memory, in service to fertility; labor; planting; grafting; raising livestock, milk, and honey—also actions of magic and creativity. Perhaps they are simply the exaltation of forces that help us to persevere and give meaning to our lives. There is nothing in our ancestral traditions to contradict this view. We are simply trying to manifest and discover a story of eternal values. Of course they are cultural in form, but they become vital forces when manifested with good will. I have known many rationalists whose so-called realism has transformed them into vain, intolerant, sterile, closed-minded tyrants. Of course, they always treat with condescension at best any sense of the sacred such as ours. To them, we have not sufficiently evolved to rid ourselves of superstition. These people are blinded by a fundamental confusion between the manipulative power of religious doctrines and the radically different power of the invitation to all humans to grow, transform, and cut loose the millstones from our necks so we can breathe again. But extremists always resemble each other, as I'm sure you know.

"When this adventure was still in its early stages, it was easy for all the conformist experts (both agronomists and ecologists) to dismiss our efforts as futile and utopian. But now their attacks have been silenced—our results speak louder than any words could. The one thing we have yet to find, however, is our place in the larger society. One thing is clear: we have no need whatsoever for your so-called economic growth. We know from experience that this idea of growth ruined our society just as it is ruining the planet right now. The coming times will be decisive. If you in the north do not succeed in freeing yourselves from the demon of growth at all costs, that demon will devour you all—not just your poor, but your rich, your politicians, your businessmen, and your financiers.

"We do not blindly reject all notions of growth and development, but these ideas must make sense within our scale, capacities, and aspirations. When we have stabilized our basic issues of food, clothing, shelter, and health for everyone, we will go further. Our own values and dreams invite us to do this—and we are always renewing our dreams. It takes only a little dreaming to feed the joy of life. Yet the modern obsession with more, bigger, further, more powerful . . . we feel this is not a true dream, for unlike a dream, it makes people more unhappy and anxious. You can see this in their faces in all the cities of abundance in the north. But of course, this doesn't mean that we've achieved some ideal. We are still struggling to overcome the effects of a long lethargy and resignation in an effort to mobilize our true strengths and recognize our true needs. We have also known the excesses of recklessness and passivity, accompanied by corruption, bloodshed, and the rule of unscrupulous tyrants. This great continent of Africa is dying, even though it is full of wealth and is underpopulated. Beneath the appearance of happy order in this village, there are still men and women who harbor in their hearts fear, envy, frustrated desires, and sometimes even violence. Like all human communities, we need to evolve toward something better, but our basic joy in living and working together is undeniable. You will find no instances of loneliness and marginality here. Perhaps there is something genetic in our sense of solidarity, for our traditional social systems have always placed the highest priority on reciprocity and community."

I had been listening attentively to Ousseini and understood all his arguments. Yet there was one subject that I had refrained from bringing up until now, perhaps out of a sense of discretion. The constantly audible presence of many children in the village, their moods ranging from outbursts of delight to bad temper, brought the question to the forefront of my mind: What about demographics and birth control? I posed this to Ousseini in the simplest way possible to Ousseini and he paused long before answering, as if emerging from a reverie.

"Of course, there are too many people on the planet, right? And along with this assertion there is the assumption, either spoken or unspoken, that there are *especially* too many people in underdeveloped countries. But is not this situation the result of the vast, instituted selfishness that we call the "world order," which we now recognize as an all-devouring disorder? In truth there is plenty of water, earth, plants, animals, minerals, energy, and good ideas to satisfy the needs of all people. It is civilization that has spoiled this wealth. After inflicting an inhuman order upon the world, those who have created this new order now want to make amends with humanitarianism—but of a kind accompanied by many perversions. I have lived in your country and seen how many people could be fed from the garbage thrown away every day by average families. Many of your dogs and cats are lavished with special food and veterinary care and the kind of love that would be the envy of countless children living on the streets of big cities. Of course these animals are innocent; they did not ask for this treatment and we can feel only compassion for the lonely and isolated people who are consoled by their presence. But does that make it a normal, acceptable situation?

"Even more important, the system of monoculture built solely around exports, often at very unfair prices; the trade wars; the droughts that ravage more and more farmland—these have led to an ongoing mass migration in all countries away from rural areas and toward big cities, which are nothing but traps of misery for most of these migrants. But you know all this. The very existence of such cities is the real source of population problems, festering like a disease.

"Those who most loudly deplore the problem of overpopulation are usually comfortably well-off themselves, with access to good social services and a pension for their old age. Among us, an old person, whether a man or a woman, is condemned to a life of misery if there are no grown children who will someday grow up to take care of him or her. For countless centuries, this was the natural order of things, and it worked. This way was in harmony with nature, and death exacted its necessary and severe tribute to maintain the balance. But in achieving control over death through medicine and other wonders, we have neglected to balance it with an understanding of what prosperity really means. If we were mindful of this, the problem of population would not be what it is.

"Libraries are full of studies and analyses showing many viable solutions, yet nothing is done; no one acts. This is the real problem. The problem lies with humanity in its entirety, not just with a few overpopulated areas. What will we do? Are we humans no more than a colony of parasites on this planet, devouring ourselves and following our own inevitable process to its certain fatal and absurd conclusion? Or can we awaken to the solutions and find the resolve to apply them?"

Even while expounding on these dark dilemmas, Ousseini's tone often had a curious lightness about it. Sometimes he would give a little snort of derision, as if to mock any indulgence in pessimism or morbid sense of doom. It was clear that this was a man whose every thought and action were grounded in love and faith in life, yet he also knew his limits and accepted them with humility.

Meanwhile, my own thoughts were returning to my world of the West and to the grim reality behind its glittering appearances. Confronted with its economic recessions, its swelling numbers of marginalized citizens, the rapid disappearance of its small farmers, a world continuing on its drunken binge of materialism and technophilia, it was clear to me that the West is now staggering, desperate to maintain the illusion of an unlivable model, continually bolstering that illusion with new illusions . . . but for how long?

One day Ousseini said to me: "Someday you will discover, just like we have, that the earth can once again become your salvation. When history swerves this far off its tracks, people rediscover the need for true, durable, indestructible values grounded in life itself, and not in all these baubles and distractions that hypnotize and excite the envy and greed of the poor. How much longer can your industries continue blindly to churn out goods that fewer and fewer people can afford? And what about your huge rates of unemployment and your endangered retirement plans and health systems? Do your people really think life can continue while cities explode in population and farmland grows to be more and more underpopulated? Will you continue to live as hostages to monstrous social and corporate structures that decide everything for you? Even what you eat is determined by vast overproduction and distribution through chains of supermarkets and gigantic, costly transportation networks. Will you continue to pay attention to your media, which always urges you to buy and consume more, hoping this will somehow fill the void of your impoverished imagination and your lack of any real satisfaction in life? Someday your people will discover that your feared recessions are actually your salvation, for they will bring you back to a true sense of life and calm the pains inflicted on you by this system. They will help to free your minds from a terrible hypnosis. You too will understand that nature offers her gifts for the profound well-being of all life, and not for the making of money. You will rediscover, as we have, how humans can make their own destiny in their own communities, which are both autonomous and interdependent. This planet has the potential to become a tapestry of communities, each woven from the seriousness and reciprocity of its own people. Such a diversity of cultures and races would enrich us all. You will help the rest of the world not by offering hypocritical aid to us, the deprived, but by helping your own people. . . . But forgive me! I am indulging in too much emotion. Let us speak now of other things. . . ."

Naturally, we spoke much of Tyemoro, whom Ousseini regarded as a great man. I told him of my apprenticeship to him in learning about Batifon culture, of my voluminous notes and tape recordings, all based on the gift of Tyemoro's prodigious memory. I also shared with Ousseini a question about Tyemoro that had long been puzzling me: How is it possible for one man to contain such a vast and precise oral memory? Is this some natural ability that we literate people have lost through centuries of reliance on the crutch of our writings? I confessed that in not understanding how it could be possible for one man to remember such a vast amount of material with such precision, I sometimes even doubted the validity of Tyemoro's information.

Then it occurred to me to invite Ousseini to listen for himself to the tape I had made of Tyemoro's long account of Ousseini's teachings. He readily accepted this offer, and I left him alone with the tape recorder and went for a long walk in and around the village.

On my return, I found him sitting in his room, silent. No doubt he had just finished listening to the tape. I sat with him and waited. Would he have comments or perhaps corrections to offer about Tyemoro's account of his own words?

He remained silent for sometime. Finally, he spoke, as if wondering to himself: "How did he do it? My talk to his people was really just a long ramble of didactic stories and anecdotes, in no particular order. But he has given my words a coherence, simplicity, and order that makes them a real teaching. He has pruned the inessential parts and gone right to the heart of the matter. You see, this shows the power of a true initiator: memory and words are taken and modeled like clay to bring out their true message. I hope you realize how lucky you are to have this."

Indeed, I knew how lucky I was, and I was grateful for the trials I had to pass through in order to gain this luck. But who among my own people would ever really understand this? For now, it would have to remain my own secret.

Clearly, Ousseini and I were now brothers, united by an intensity of shared feeling. We promised each other always to keep in touch and work in our different worlds toward a common aspiration. I felt certain now that I would often come back here.

Nine months after my return to Europe, I received a letter from Ousseini informing me of the death of Tyemoro. As I opened the envelope bearing the marks of a long voyage and covered with vibrant, colorful stamps, I had no premonition of the pain that its contents held for me. It was written on a simple piece of notebook paper:

"He called me to his bedside. There was a profound peace in that bare, simple room of a poor man. All the villagers were waiting outside, in dignity and silence. He asked me to promise to continue to take care of his people and their land. He also said, 'Tell my white son. Tell him he is always in my heart, and that I have confidence in him.'

"I left him with his own people for a while. He spoke with them for a long time, then he called me back: 'I have advised them to obey you, for you are a human seed. Continue your work, but know this: You will belong less and less to yourself, for you are the good that belongs to all. Many others must also become the good that belongs to all.' After a long silence, he continued: 'Now that I have no more needs, I would ask of you one last favor. You may find this strange, but I'll risk it—it is my last risk. Maybe people will say the old man went a bit crazy at the end . . . so what? Here is my wish: When you have placed my body in its last home in the earth, curled again like a fetus, I would like for you to spread some of that good black earth over my body. I am leaving this world without really understanding what this food means. I only know that it is a key to the reconciliation between humans and Mother Earth. There may be other ways, but this one opens all of them.'"

An Interview
with Pierre Rabhi

by Joseph Rowe

he drive to his farm, Montchamp, takes us into country that
becomes more and more rocky and arid. This is the Ardèche, one
of the poorest regions of southern France, but one that is studded with
places of great beauty. Less than an hour's drive downhill from here is
the breathtaking Pont d'Arc, a natural stone bridge straddling one of the
most splendid stretches of the Ardèche River.

The roads become rougher as we climb, and the trees have now
almost entirely disappeared, replaced by granite outcroppings and dry
scrub brush. Though this country is normally arid, we have learned that
it is currently suffering a very intense and abnormal drought. Finally, we
make the final steep climb up to the plateau of Montchamp. We park
the car and note the large, centuries-old stone farmhouse that dominates
the scene. The most striking sight, however, is the sudden appearance
of greenery everywhere and groves of trees, many of them fruit trees in
bloom. If I did not know better, I might think that we had crossed into
a different bioregion. This flourishing oasis of fertility is the creation of
Pierre Rabhi, a largely self-educated agro-ecologist, writer, and international consultant.

He warmly greets my wife and me outside and leads us toward an
entrance to his house. We sit at a table near a large window with a

pleasant view of the gardens and fields outside. Pierre is a short, calm though energetic man with graying, crinkly hair and a mustache and goatee. There is something of the African sage in his glittering, deep-set, intensely alive eyes. Immediately, we feel at ease with him. Though French is not his mother tongue, he speaks it perfectly, with poetic eloquence and precision. At one point we pause in our discussion as he goes off to brew and serve us the delicious, sweet mint tea, which is traditional in the Maghreb countries of North Africa.

Joseph Rowe: You have coined a new term, agro-ecology. What does this mean and how does it differ from what is known as organic farming?

Pierre Rabhi: Agro-ecology is a much larger category than organic farming. In 1981, I was asked to help the farmers in Burkina Faso. In France, I am considered an organic farmer, but in Africa, I knew that would not be enough. I was faced with the harsh reality of a land devastated by drought and desertification, much of it due to the destruction of forests by corporate interests. So I could not help the farmers without also trying to restore the ecological balance by planting many trees, establishing sound water conservation and management practices, and using numerous other techniques that would begin to repair the damage done to the local ecology. To me, ecological balance must include appropriate agricultural practices, not just protection of wildlife, and vice-versa.

JR: Farmers are part of the ecology, after all. . . .

PR: Of course they are. And another reason I prefer to describe myself as an agro-ecologist is that this term avoids all the disputes that arise as to who is certifiably organic and who is not. Agro-ecology does not have to answer to any rules about whether farming practices are 100 percent organic. What is important is what is good for the ecological balance of the whole region and for a truly sustainable agriculture in which the farmers' first priority is to assure their own food supply. This naturally consists essentially of practices that are considered organic and sometimes biodynamic. But these labels are unimportant. These farmers are not trying to qualify to sell produce to urban health-food consumers, they're trying to survive and save their land in the process. And even if they do reach a point of abundance where they have cash crops, the first priority of the agro-ecological approach must always remain feeding themselves properly, feeding the land properly, and maintaining the ecological balance.

JR: One thing I've noticed, both in the United States and in Europe, is that fruits and vegetables that come from huge, industrial organic farms often have surprisingly little taste. On the other hand, the same kind of fruit from a small farmer in your area who is not necessarily organic often has a far better taste.

PR: Yes, this is a common phenomenon. You see, the essential thing is our attitude toward the land. Technique is secondary. I maintain that when people lack a sense of the sacred with regard to the earth they farm, the people and creatures who share it with them, and those who eat the food it produces, then there is something fundamentally wrong. Their farming practices may be impeccably organic—and certainly this is better than chemical farming—but they are still treating the land as a factory. Getting it to produce as much as possible in order to maximize profit is their real motivation. And as you say, you can often taste the result in their produce, which is harvested as prematurely as possible so that it can be stored longer and shipped greater distances.

JR: In your introduction to this book you define the sacred as "a sense of humility in which gratitude, knowledge, wonder, respect, and mystery all come together to inspire and enlighten our actions."

PR: Well, I'm happy if those words are helpful, but of course they are not a definition. It is important that the sacred not be imprisoned by words. Unless it permeates our most concrete, everyday experience, it will be lost. By this, I mean, for instance, simple awareness of the sacrifice of other forms of life so that ours may continue—and the gratitude that always accompanies this awareness, if it is true. This sacrifice is present, whether we acknowledge it or not, every time we eat a meal. It also means an awareness of our place in the great circle of life and death on this earth and a sense of responsibility for what we are doing to this circle.

The overwhelming lack of a sense of the sacred in modern people is not a reflection of some superior knowledge we have and certainly has nothing to do with any sort of progress. On the contrary, it represents a regression and a loss.

We humans are creatures with a special endowment of intelligence and consciousness, but instead of using this gift to deepen our sense of wonder and beauty and make this world even more wonderful and beautiful, we have chosen the path of domination and subjugation of nature. As soon as we made this choice, we began to harm ourselves, for the attempt to separate ourselves from nature is a destructive illusion.

JR: It would seem that so-called rational discourse, whether in agriculture, business, politics, or science, always relegates this dimension of wonder and beauty to mere subjectivity.

PR: Yes, but many of the greatest scientists have long abandoned that way of thinking. Such rationalism is not the true foundation of science, though unfortunately, it is still the dominant mode of thought in the modern world. By the way, I am not using the word *rationalism* to mean "placing a high value on reason." Rather, I mean the dualistic separation of reason from beauty and wonder. In agro-ecology, we are obliged to be quite rational, scientific, and technical. But this is inseparable from the beauty and reverence that are just as fundamental to our work. All of these are aspects of what I call the sense of the sacred.

In my life, I have been very inspired by reading the dialogues between Krishnamurti and the physicist David Bohm. Here (especially in their dialogue about time), a great scientist and a great mystic from very different paths in life come together in the realization that the observer cannot be separated from the observed. On many levels—not just the level of quantum physics—the observer cannot help but be a part of what he observes, interacting with it and changing it. This often works through hidden assumptions and beliefs. In their dialogues, Krishnamurti and Bohm show, among other things, that the relegation of beauty to a merely subjective domain is really quite irrational.

JR: Now you seem to have turned the word around!

PR: (laughing) Yes, and I'm glad to do so. I wouldn't care to be labeled an *irrationalist* any more than a rationalist.

JR: In your book, there is a powerful leitmotif that I interpret as a kind of diagnosis of the fundamental sickness of modernism: its systematic rejection or avoidance of the sacred. In the United States, this has become a very reactive political issue, with many people wanting to restore religion to what they believe is its rightful place in politics. But for some reason, this seems to be happening only on the extreme, fundamentalist right.

PR: Well, many of those people also seem to have lost their sense of the sacred, no matter how loudly they preach or how often they go to church. Please forgive the caricature, but often my image of U.S. politics is that of a man with his pockets stuffed with dollars, brandishing a Bible in one hand and a gun in the other. . . .

No, living with a sense of the sacred is not a function of whether or not a person holds specific religious beliefs. And it's true that the sacred must have a central place in politics, provided it is not linked to acceptance of a religious or any other ideology.

JR: Could you say more about how this might come about?

PR: You'd think that a sense of the sacred would find its most obvious, natural expression in all public discourse oriented toward ecology, sustainability, and environmental conservation. Yet somehow, it very rarely does. To me, this explains the fundamental, persistent failure of all the Green parties, in spite of their minor successes. They can quite thoroughly inform you about what is wrong, but they have no vision that engages true enthusiasm. Of course, we must speak of what is wrong, of the terrible damage we are doing to the earth. But if we do so without a sense of the sacred and without the courage to let the sacred show, let it be heard and felt by others, we ultimately fall prey to negative emotions. This is never going to bring about any major change.

JR: Yes, it is as if the Greens, like other political parties, are actually embarrassed by too much openhearted talk of the sacredness of the earth and of nature.

PR: This is partly because of their fear of the sacred, but it is also because we have all heard too much chatter about the sacred, which inevitably leads to some form of proselytism. I am not advocating any sort of proselytism of the sacred in politics. I am advocating an open and unashamed *expression* of the sacred in politics. The sacred is unspeakable, beyond words—yet it can readily be felt in and between our words, if we allow it to simply dwell in us and guide us. Whenever I talk about ecology in any public situation, I try to share with people the same kind of awe and mystery I feel in my garden when I contemplate a seed—a tiny seed that could someday become a tree that feeds many people. I refuse to consider this reverence and wonder as some sort of private feeling that is out of place in politics. On the contrary, public life desperately needs more of this kind of expression of feeling. But it will demand a radically different approach to politics, beyond all the right/left games and other dualisms.

JR: Was this what you were trying to accomplish in your candidacy for president of France in 2002?

PR: At first, I was very reluctant to get involved in such a campaign. I have no ambitions at all in that area. But some friends of mine brought gentle pressure to bear on me and convinced me that it could be useful. What they finally convinced me of was that it could be a good testing ground to demonstrate that a very different approach to politics *is* possible. And our campaign turned out to be quite provocative, actually.

JR: How so?

PR: For one thing, we rigorously refrained from any sort of criticism of opponents, any blame-oriented polemic, or any sort of adversarial discourse that would situate us within the right/left political spectrum. We did express vehement and fundamental opposition to certain policies, but we never indulged in attacking individuals or even other parties.

This, plus our willingness to express—not just talk about—our sense of the sacredness of the earth, turned out to be quite provocative. It was as if people—including some who are normally cynical and apathetic when it comes to politics—were dying of thirst, longing to hear something like this. In spite of our total lack of professionalism in organizing a campaign, we had packed houses for all our appearances, and for such a tiny campaign with almost no financing, very surprising results in collecting signatures. Without even trying, we wound up collecting more signatures than the Green candidates. Of course, it was not nearly enough to get us on the ballot, but it was a sign of this thirst I am referring to.

Furthermore, the quality of the debates and discussions was extraordinary. People really listened to each other, even when they disagreed, and often came forward to speak with passion, clarity, and honesty about the real issues facing us. Some of them had the courage to speak of the joylessness and even despair they feel in living in a consumerist, materialistic society. This was unlike anything you normally see in political meetings. It seemed to awaken a communion of feeling and intelligence far beyond my own offerings. I often like to refer to this collective emergence as an "insurrection of consciences."

But the nonblaming attitude is important. All the models of oppressors-and-oppressed systematically overlook the potential for oppression hidden in the hearts of the oppressed. History has demonstrated this over and over. We have also found that this new approach is very much related to a renaissance of respect in politics for the nurturing, feminine principle. A major symptom of the destructive imbalance of our world is the overmasculinization of our societies, with their exaggerated worship of competition. A better balance doesn't just mean more women running for office, it means men and women acting together, expressing this balance and partnership between the masculine and feminine principles in all of us. One of the great gifts of the feminine principle is its ability to reveal the sacred as implicit in a powerful, ancient way that can be felt by everyone, without the need for doctrines.

JR: So there can be no significant political or social change without a change in people's hearts and minds on an individual level. But this seems to lead to a chicken-or-egg kind of situation. There can be no social change without individual change, yet individuals need support-ive social structures to help them to change.

PR: It's true, this all must happen together. But I feel that our campaign effort was at least a step in that direction. What other political party—at least in France—has ever dared to mention any sort of need for individuals to change their lives, much less undergo what amounts to a personal transformation? They never even hint at such a thing, probably for fear of angering or frightening their voters. Yet it is central to our approach. And people have shown that, far from being threatened or offended by it, they are eager to hear more about it and to discuss it.

So at least this was a first step. We essentially saw it as kind of litmus test, to see if people were, as we hoped and suspected, interested enough to support an approach to political and social change that goes beyond ideology, beyond all the models of *good us* vs. *bad them,* and one that is not afraid to address people's own individual yearning, that sense that they must change their lives. This test succeeded far beyond our expectations.

JR: When I told an American friend about the project of translating this book and a little about you, I described you as a French agro-ecologist of Arab origin. Her reaction was: "An Arab ecologist? Thank God there's at least one!"

PR: Well, that's an understandable reaction. We hear very little about ecological activism in the Arab world. It's unfortunate. Too often, our notions of Arabs are stereotyped by media images from the oil-rich countries of the East and the torments of war.

But the truth is, I can't exactly be described as an Arab. I was born and raised in a Muslim, Arabic-speaking society in Algeria, but I am really a North African Saharan. Our family has more Peul, Berber, and Tuareg blood than Arab. You can see it in my features. Actually, our family is such a mixture that I have some relatives who have chocolate-colored skin, and others who are almost white—but these are just personal details. My destiny has been that of a cultural hybrid: I was later adopted by a French family and baptized as a Christian. Perhaps it is this hybrid destiny that has helped me to refuse to identify with any particular culture, religion, or ideology.

JR: I get a sense that there is a lot of you in your character Ousseini.

PR: Actually, the character of Ousseini was inspired by a remarkable man I met. He came from a poor African country and was a gifted intellectual who had spent many years studying in Europe. He had several doctorates—this man was covered with diplomas! And so was his wife. They came to visit me here, and we had a very good exchange. He was sitting right there, in the same chair you are, when he told me of his intention to return to his country and his quandary about how he could really help his people. He knew he could not help the wretched situation there by living in a big city and trying to obtain (probably nonexistent) funds for setting up some sort of laboratory or research center, which he was fully qualified to do. So he returned to his native village with his wife and announced that he was going to become a farmer. This caused total shock and consternation. His own people refused him land, and he had to set up in unwanted lands that his people believed were inhabited by hostile spirits. This man's courageous example was the main inspiration for Ousseini. But of course you're right, there's also something of me in him, especially in the practical details of his teaching.

Also, I had long ago had an experience that put me in natural sympathy with my African friend. When I first moved to this land in 1960, I was able to buy the house and acreage for an amazingly low price because it was considered unfarmable. The local people thought I was crazy to try and were sure I would fail and leave. But over time and through a long initiation of my own, partly inspired by the principles of biodynamic farming, I finally began to get good results. Those who saw the results wanted to learn more. That's more or less how it all began.

Pierre Rabhi,
Farmer without Borders

Note: The following is an extract from an article by Christian de Brie, "Pierre Rabhi: Paysan sans frontières," in *Le Monde Diplomatique*, July 1995.

With an implacable logic, intensive agriculture is imposing its destructive model of development all over the world. The devastation it causes is less and less compensated by its advantages. "In the third world," as François de Ravignan points out, "the damage is often more spectacular, because of the brutal subjugation of peasants to market forces, which leaves them with no choice but to deplete their soil through overgrazing, water pollution, erosion, destruction of forests, and the loss of aquifers without any means of renewing it. Meanwhile, populations continue to grow, with increasing needs for food."

In Europe and North America, farmers must remain competitive in order to survive. Hence they must continue to increase their production, plowing ever-larger, more level, deforested fields to be able to accommodate more and more powerful machines designed to assure maximum production. Farmers are caught in a vicious circle in which monoculture and highly specialized grains produce species that are more and more fragile and less resistant. More fertilizer, more pesticides, and more plant treatments against parasites are required. Meanwhile, parasites are becoming increasingly resistant to these dangerous and costly treatments. Farmers who succeed do so at the expense of their less efficient neighbors. Agronomic research is fundamentally oriented toward greater and greater productivity, by any means possible. This study includes genetic and other biotechnological manipulations whose purpose is to enable multinational corporations to increase their market share, their profits, and their control over the food supply.

Far from serving agriculture, industry and finance have seized control of it in order to maximize their profits. The farmer has become a captive client for mechanical and chemical products whose costs keep increasing. Falling totally in the thrall of banks, they incur lifetime debts that now amount to colossal figures (over 200 billion dollars in the United States, a country where the public cost of creating a new job in agriculture is now several hundred thousand dollars, more than in any other sector of employment).

Agricultural produce has become a gambling chip played in speculative agribusiness markets in which farmers have no influence and to which they have no access. The price they receive for their produce barely covers its real cost and is often well below this figure when public subsidies are involved. After packaging and advertising, this produce is then sold to consumers at three to ten times the price in supermarket chains who profit further by squeezing out local agricultural economies.

In this scenario, every plant and every animal is being progressively reduced to mere computer data, which directs everything the farmer manages even as authentic farming knowledge and experience are lost. Money and resources are funneled into overproduction, which then necessitates greater spending on subsidies. This is a model of development that wastes as much as it produces; destroys and diminishes humans and their knowledge; impoverishes genetic diversity and food quality; and ruins soil fertility, aquifers, and nonrenewable energy and mineral resources.

When all the true and hidden costs are calculated, this model of agriculture is the least efficient and least sustainable ever devised. Yet it is this model that markets impose everywhere, with the support of international organizations in the United States and Europe. Even now, experts are preparing to reduce the number of rural food producers to less than 3 percent of the working population. Globalization of agricultural markets and the practice of dumping have dismantled traditional food production systems all over the planet. This is especially true in Africa, where the huge cost of incessant international shipping and transport have created artificial shortages and poverty. As a result, ever-growing populations of people have lost the ability to feed themselves, becoming dependent on imports—if they can pay for them—or, in extreme situations, on humanitarian aid.

Confronted with this power structure, small farmers of both the north and south face essentially the same calamities: loss of autonomy, rural exodus, environmental degradation, and deterioration of the quality of life. In time, a world food crisis is not that unlikely, for agricultural war is already a reality. Yet the real resources are far more than sufficient for the needs of all human beings on this planet. The productivist model still dominates, with its powerful illusion of efficiency, yet its credibility is steadily eroding, especially in Africa, where the impossibility of falling in step with the global marketplace is so obvious that people are being forced to question the model and try new approaches.

A dozen or so students have come to a workshop on tropical agro-ecology near Montpellier, in France, sponsored by the Carrefour International d'Echanges de Pratiques Appliqués au Développement (CIEPAD).* They come from Burkina Faso, Benin, Cameroon, Burundi, Brazil, and New Caledonia. Some of them are agronomic technicians, while others are students on scholarships offered by NGOs or governments. The experiential part of the workshop will take place in Burkina Faso. What this experience offers is not mere technical recipes and models, but tools for effecting a deep analysis, evaluation, and experimentation that aims at a truly sustainable agriculture—in other words, one that respects human beings and environments while highlighting problems and issues faced by farmers in both Europe and elsewhere, in both temperate and tropical zones.

---

*[CIEPAD no longer exists as such and has been replaced by the organization Terre et Humanisme. See http://terrhumanisme.free.fr/ for more information relating to Pierre Rabhi and his work.—*Trans.*]

The priority of the agro-ecological approach is for people to be able to feed themselves through respect and effective use of their own local resources. Production must take second place to values that link agriculture with environmental protection. Emphasis is placed on organic composting, maintenance of soil fertility, and plant treatments that are as natural and biodegradable as possible. Favored are traditional species and varieties of plants and animals that have been adapted to local conditions and skills. Optimal management of water resources is taught, and great caution is taken to minimize use of equipment that demands costly fossil fuel. A high priority is placed on anti-erosion (terracing, dikes, vegetation barriers) and reforestation. Finally, there is a policy of rehabilitation of traditional skills, wisdom, and knowledge adapted to ecological stewardship of the area. This is an educational process that is global in its implications and is relevant everywhere.

CIEPAD has also developed a program that addresses the need to increase awareness and knowledge of agro-ecological principles among educators and students in the north and linked to concrete programs of sustainable development in the south, working with local leaders in places such as Senegal, Togo, Benin, Burkina Faso, Mauritania, North Africa, and other countries.

In Tunisia, for example, CIEPAD is working in the Gulf of Gabes on a project to restore an oasis threatened by desertification. In the Maghreb, some half-million acres of oases are now threatened by productivist technology in flagrant disregard of the fact that these ancient, human-created micro-ecologies depend on a fragile balance that must follow extremely strict and precise rules in order to survive.

In Palestine there is a project at Falaniah, west of Napluz, which involves training farmers in agricultural practices that are less dependent on intensive irrigation and on Israeli agribusiness markets. This project is also linked to a women's group that cultivates aromatic and medicinal plants as a source of revenue and autonomy.

Another CIEPAD project is an experiment in an optimal agricultural installation model [MOIA is the French acronym], which works to enable a family of four to sustain itself on a small, fully organic truck farm. The aim for the family is full food self-sufficiency with a surplus that can be sold in local markets. This model include a transportable ecological habitation* and is possible with a very modest loan that is well within the means of small credit cooperatives.

An overall goal is the repopulation of abandoned rural lands through a sustainable productive activity that includes both autonomy and time for other activities such as the healing and restoration of human relations and services in the affected areas. Another aspect of this program addresses the current dire need for a saner relation between overpopulated cities and abandoned rural areas—one that works toward a new kind of rural-urban society that is more compatible with a sharing of work and life than the existing model that includes the typical urban quest for "quality of life" shaped by the dehumanized values of a consumerist society. This project is also concerned with reaching out to people who find themselves on the margins of society with few resources—those who are appropriate candidates for CIEPAD training in setting up a MOIA.

The underlying goal is always the same: to demonstrate in the most rigorous and convincing way that people who are not farmers in the traditional sense of the term can successfully take up a life of farming while rediscovering autonomy, community, and quality of life. This is possible both in the north and the south. It is happening now, a quiet, nonmediated movement, but a vital one. All over Europe, especially in France, we can observe new rural communities forming and a small but discernible growth in rural population. When visiting these communities, we may well have the impression that creative imagination is far more alive and well in the rural countryside than at the conferences of the experts.

Pierre Rabhi, founder of CIEPAD, the son of an oasis blacksmith, was born and raised in the southern Sahara of Algeria. He was educated in France and had long been the bearer of a dual Franco-Maghreb culture before he set up in the southern Ardèche, living the frugal life of a simple farmer. In his own journeys between South and North, he has developed an ethic of development based on "the cultivation and appreciation of resources available to local communities in their own lands. Local production and consumption must become the foundation of true internationalism." His relationship to the land is not one of productivist

---

*[A "transportable ecological habitation" refers to a kind of sophisticated tent, though far warmer, more comfortable, and more habitable than a traditional structure of this kind.— *Trans.*]

exploitation, for the soil is a living being, a nourishing mother, and must be treated as such.

He has applied the practical aspects of his philosophy with extraordinary success in the arid land of the Cevennes. Only after this achievement did he begin to teach others. This transmission led to the creation of a training center in the Sahel, in Burkina Faso. In the beginning, it was supported by the active assistance of President Sankara, and afterward by the CIEPAD foundation. A down-to-earth visionary, an Ardèche peasant rooted in his land, he is also a farmer without borders. He manifests a happiness that arises from voluntary simplicity and advocates the creation of new human oases—ones that are not closed and insular, but open to others. In a world where "everything has a price, but nothing has value," this sobriety comes as a form of liberation. As he himself says, "it is not the gross domestic product that inspires human dreams, but a sense of sharing, and equality."

Books of Related Interest

The Rebirth of Nature
The Greening of Science and God
by Rupert Sheldrake

The Presence of the Past
Morphic Resonance and the Habits of Nature
by Rupert Sheldrake

The Secret Teachings of Plants
The Intelligence of the Heart in the Direct Perception of Nature
by Stephen Harrod Buhner

The Universe Is a Green Dragon
A Cosmic Creation Story
by Brian Swimme, Ph.D.

Green Psychology
Transforming Our Relationship to the Earth
by Ralph Metzner, Ph.D.

Science, Soul, and the Spirit of Nature
Leading Thinkers on the Restoration of Man and Creation
by Irene van Lippe-Biesterfeld with Jessica van Tijn

Science and the Akashic Field
An Integral Theory of Everything
by Ervin Laszlo

The Nature of Things
The Secret Life of Inanimate Objects
by Lyall Watson

Inner Traditions • Bear & Company
P.O. Box 388 • Rochester, VT 05767 • 1-800-246-8648
www.InnerTraditions.com

Or contact your local bookseller